Managing Supply Chain Risk and Vulnerability

Teresa Wu · Jennifer Blackhurst
Editors

Managing Supply Chain Risk and Vulnerability

Tools and Methods for Supply Chain Decision Makers

Springer

Editors
Assoc. Prof. Teresa Wu
Department of Industrial Engineering
Ira A. Fulton School of Engineering
Arizona State University
P.O. Box 875906
Tempe, AZ 85287-5906
USA
teresa.wu@asu.edu

Assoc. Prof. Jennifer Blackhurst
Department of Logistics, Operations
and Management Information Studies
College of Business
Iowa State University
Ames, IA 50011
USA
jvblackh@iastate.edu

ISBN 978-1-4471-5783-0 ISBN 978-1-84882-634-2 (eBook)
DOI 10.1007/978-1-84882-634-2
Springer Dordrecht Heidelberg London New York

British Library Cataloguing in Publication Data
A catalogue record for this book is available from the British Library

Cover design: eStudioCalamar, Figueres/Berlin

Printed on acid-free paper

Springer is part of Springer Science+Business Media (www.springer.com)

Contents

List of Contributors

Jennifer Blackhurst
Department of Logistics, Operations and MIS College of Business, Iowa State University, Ames, IA 50011, USA

Clare Brindley
Nottingham Business School, Nottingham Trent University, UK,
e-mail: clare.brindley@ntu.ac.uk

Ying-Ju Chen
University of California, Berkeley and University of Texas, Austin

Jose M. Cruz
Department of Operations and Information Management, School of Business, University of Connecticut, Storrs, CT 06269, e-mail: jcruz@business.uconn.edu

Shoumen Datta
Engineering Systems Division, Department of Civil and Environmental Engineering, MIT Forum for Supply Chain Innovation, School of Engineering, Massachusetts Institute of Technology, 77 Massachusetts Avenue,
Cambridge MA 02139

Jin Dong
IBM China Research Lab, Zhongguancun Software Park, Beijing 100193,
P.R. China

June Dong
Department of Marketing and Management, School of Business, State University of New York at Oswego, Oswego, New York, 13126

Pat Doody
Department of Mathematics and Computing, Centre for Innovation in Distributed Systems, Institute of Technology Tralee, County Kerry, Ireland

Mohd. Nishat Faisal
Institute of Management Technology Dubai, P.O. Box 345006, Dubai International Academic City, Dubai, United Arab Emirates, e-mail: nishat786@yahoo.com

Mark Goh
NUS Business School, The Logistics Institute – Asia Pacific, National University of Singapore, 7 Engineering Drive 1, Singapore 117543, Singapore, e-mail: mark_goh@nus.edu.sg

Don P. Graham
INNOVEX LLC

Olli-Pekka Hilmola
Lappeenranta University of Technology, Kouvola Research Unit, Prikaatintie 9, FIN-45100 Kouvola, Finland

Chun-Che Huang
Department of Information Management, National Chi Nan University, No. 1, University Road, Pu-Li, Na-Tau, 545, Taiwan, R. O. C.

Fanwen Meng
NUS Business School, The Logistics Institute – Asia Pacific, National University of Singapore, 7 Engineering Drive 1, Singapore 117543, Singapore, e-mail: tlimf@nus.edu.sg

Anna Nagurney
Department of Finance and Operations Management, Isenberg School of Management, University of Massachusetts, Amherst, Massachusetts 01003

Qiang Qiang
Department of Finance and Operations Management, Isenberg School of Management, University of Massachusetts, Amherst, Massachusetts 01003

Bob Ritchie
Lancashire Business School, University of Central Lancashire, e-mail: rritchie@uclan.ac.uk

Nikhil Sagar
Retail Inventory Management, OfficeMax Inc, 263 Shuman Boulevard, Naperville, IL 60563

Sridhar Seshadri
University of California, Berkeley and University of Texas, Austin

Reuben Slone
OfficeMax Inc, 263 Shuman Boulevard, Naperville, IL 60563

ManMohan S. Sodhi
Cass Business School, City University of London, e-mail: m.sodhi@city.ac.uk

Christopher S. Tang
UCLA Anderson School, e-mail: ctang@anderson.ucla.edu

Tzu-Liang (Bill) Tseng
Department of Industrial Engineering, University of Texas at El Paso, 500 W. University Ave., El Paso, TX 79968, USA

Wei Wang
IBM China Research Lab, Zhongguancun Software Park, Beijing 100193, P.R. China

Teresa Wu
Department of Industrial Engineering, College of Engineering, Arizona State University, Tempe, AZ 85287, USA

Chapter 1
Book Introduction

Jennifer Blackhurst and Teresa Wu

A supply chain is a network of entities such as manufacturer, suppliers and distributors working together to transform goods from raw material to final product while moving them to the end customer. Effective supply chain management is a critical component of a firm's ability to fill consumer demand, regardless of the industry. Supply chain performance may be decreased by disruptive events occurring in the supply chain system. Supply chain disruptions are "unplanned events that may occur in the supply chain which might affect the normal or expected flow of materials and components" (Svensson, 2000).

Managing the risk of these events occurring in the supply chain has become known as supply chain risk management (SCRM) and may be defined as "the management of supply chain risks through coordination or collaboration among the supply chain partners so as to ensure profitability and continuity" (Tang, 2006). SCRM has recently raised the attention of both academics and practitioners and is the focus of this book.

One needs only to check the daily news to hear about disruptive events affecting supply chain performance. Some of the better known examples include the 2002 longshoreman union strike at a U.S. West Coast port and affected supply chain performance for up to 6 months for some firms (Cavinato, 2004) or the 2000 lightning bolt at a Philips semiconductor plant in New Mexico which resulted in a small fire, destroying millions of chips and ultimately their customers Nokia and Ericsson (Latour, 2001).

Because a supply chain disruption can potentially be so harmful and costly, there has been a recent surge in interest and publications in the area of SCRM – from academics and practitioners alike – regarding supply chain disruptions and related issues. *The purpose of this book will be to present tools and techniques for decision making related to supply chain risk.*

In general, a supply risk management process consists of four components:

(1) Risk identification; (2) Risk assessment; (3) Risk management decisions and implementation; and (4) Risk monitoring (Hallikas et al., 2004). In the risk identification step, risks facing the firm's supply chain are identified. Exemplary re-

search in this step may be found in Chopra and Sodhi (2004) where general risks in the supply chain are categorized and discussed. The risk assessment step involves understanding the impact of risks facing the supply chain. Examples of previous risk assessment work includes that by Zsidisin et al. (2004) and Hallikas et al. (2004). Assessing risk is a complicated step and can help a firm to prioritize which risks will affect the vulnerability of a supply chain. Risk management decision making requires supply chain managers to decide which mitigation strategies should be employed and where scarce resources may be allocated. Certainly, these are by no means easy decisions with many aspects and factors affecting these decisions. Finally, risk monitoring includes monitoring risks over time.

In this book, the chapters presents tools and methods to assist supply chain managers and researchers alike in the tasks of *risk impact and assessment* and *decision making in the supply chain considering risk*. We feel these areas are the most difficult steps of the supply chain risk management process. Therefore, the book is divided into two main sections:

1. Understanding and Assessing Risk in the Supply Chain
2. Decision Making and Risk Mitigation in the Supply Chain

These sections consist of 10 chapters from world class researchers from around the world and employ a variety of methods to address timely issues in SCRM. The methods include industry-based cases and illustrative example, fuzzy logic and analytical hierarchy process (AHP), simulation, stochastic models and knowledge management protocols. The use of this broad array of methods strengthens the contribution of this book as a whole and provides a wide range of tools and techniques for both researchers and academics alike. The method employed is chosen because it best fits the problem or issue being addressed.

The two sections are described as follows:

Section I: Understanding and Assessing Risk in the Supply Chain.

This section presents four insightful chapters relating to developing an overall understanding of risk and its relationship to supply chain performance, investigating the relationship between response time and disruption impact, assessing and prioritizing risks, and assessing supply chain resilience.

In Chap. 2, "Effective Management of Supply Chains: Risk and Performance" by Bob Ritchie and Clare Brindley the authors present an overview of SCRM and its importance to managing overall business performance. The chapter seeks to better understand the relationship between risk and performance by exploring sources, drivers, consequences, and management responses. The chapter culminates with guidelines for selecting appropriate strategies for particular drivers to maximize supply chain performance. This chapter gives the reader a high level understanding of SCRM.

The next chapter provides a more detailed level of granularity in understanding risk in supply chains. In Chap. 3, "Managing Supply Chain Disruptions via Time-

Based Risk Management" by ManMohan S. Sodhi and Christopher S. Tang a time based risk mitigation concept is developed. The promise of such a concept is to reduce the impact of rare but disruptive supply chain risk events. The authors discuss the concept that if a firm can shorten response time, the impact if the disruption is also reduced.

In Chap. 4, "Prioritization of Risks in Supply Chains of Small and Medium Enterprise (SMEs) Clusters Using Fuzzy-AHP Approach" by Mohd Nishat Faisal. The chapter focuses on supply chain failures for small to medium enterprises who may have limited resources and lack of adequate risk planning tools. The chapter introduces a Fuzzy-AHP based framework to prioritize supply chain risk. Such an approach shows promise to help firms develop strategies for managing risks in accordance with their importance.

Finally, a simulation framework is presented to better understand supply chain resiliency. In Chap. 5, "A Generalized Simulation Framework for Responsive Supply Network Management" by Jin Dong, Wei Wang and Teresa Wu a simulation tool is developed to assess the resilience of a supply network to a disruption. The chapter was created with the assistance of the IBM China Research Lab.

Section II: Decision Making and Risk Mitigation in the Supply Chain.

The first section has helped the reader to better understand and assess risk in the supply chain. From a managerial persepctive, now that supply chain risks are better understood, tools and methods are needed to assist in decision making. This section presents six chapters with tools and methods for assisting with decision making and risk mitigation in the supply chain. The chapters reflect the dizzying array of factors that need to be considered in supply chain risk decision making including supply-side and demand-side risk as well as risk attitude; contracts in the supply chain; the impact of forecasting, supply chain structure and operational policies; logistics factors and uncertainty; the impact and effect of relationships in the supply chain; and supply chain knowledge considerations. Each chapter presents a novel approach to decision making and considers certain factors.

In this section the first three chapters (6–8) investigate uncertainty and risk in a supply chain from a network perspective. Stochastic programming is a popular and powerful tool for evaluating supply chain design. Chaps. 6, 7 and 8 use this approach.

First, a network model is developed with various risks on the supply and demand side. In Chap. 6, "Modeling of Supply Chain Risk Under Disruptions with Performance Measurement and Robustness Analysis" by Qiang Qiang, Anna Nagurney and June Dong develops a multi-tier supply chain network game-theoretic model which considers, both supply-side and demand-side risks as well as uncertainty in costs such as transportation. Additionally, attitude toward risk is incorporated and a weighted supply chain performance and robustness measure is developed. This chapter is quite unique in that *both* supply and demand side risks are considered.

Relationships between the nodes in a supply chain are studied in the next chapter. In Chap. 7, "The Effects of Network Relationships on Global Supply Chain Vulnerability" by Jose M. Cruz different levels of a relationship are assumed to influence costs and risks for supply chain decision makers. A game theoretic model is developed to study different relationships among the supply chain players and how that is associated with supply chain vulnerability.

Next, the stochastic programming method is extended by adding conditional value at risk to achieve optimality. In Chap. 8, "A Stochastic Model for Supply Chain Risk Management Using Conditional Value at Risk" by Mark Goh and Fanwen Meng a stochastic programming formulation is developed using a measure termed the conditional value at risk. The focus of the chapter is on logistics-focused problem decision making where uncertainty exists.

The next two chapters in the book recognize that risks and mitigation efforts in supply chains may be much more specific in nature.

In Chap. 9, "Risk Intermediation in Supply Chains" by Sridhar Seshadri and Ying Ju Chen also explore supply chain decision making under uncertainty from the perspective of the seller distributor in the context of contract design. The chapter presents two models which can be used to create contracts where the individual retailer has different degrees of risk aversion.

In Chap. 10, "Forecasting and Risk Analysis in Supply Chain Management: Garch Proof of Concept" by Shoumen Datta, Don P. Graham, Nikhil Sagar, Pat Doody, Reuben Sloan and Olli-Pekka Hilmola advanced forecasting tools are explored to decision making in supply chain scenarios. In particular, the impact on demand amplification is investigated. Results of the chapter indicate that these advanced forecasting methods may be useful but supply chain structure and operations policies as well as data availability must be considered.

We conclude the book with a general framework which presents a method for knowledge representation in a supply chain risk context. In Chap. 11, "Supply Chain Risk Management: Annotation of Knowledge Using Semi-Structured Knowledge Model" by Chun-Che Huang and Bill Tseng develops a semi-structured knowledge model which is used to represent knowledge in the supply chain. Such a model may be used to assist in decision making related to supply chain risk.

Our goal in developing the book was to solicit contributions from top SCRM researchers around the world to develop a text that both practitioners and students can use to better *understand* and *manage* supply chain risk. We are delighted with the insightful contributions that now form this book. We look forward to the book helping to spur and contribute to interesting and insightful research as well as developing tools and methods to help supply chain managers effectively manage and mitigate risk in today's complex global supply chains.

We wish to thank each of our contributors as well as the editorial team at Springer for the opportunity to edit this interesting and insightful book.

References

Cavinato, J.L. (2004), "Supply chain logistics risk", *International Journal of Physical Distribution & Logistics Management*, Vol. 34 No. 5, pp. 383–387

Chopra, S. and Sodhi, M. (2004), "Managing risk to avoid supply-chain breakdown", *MIT Sloan Management Review*, Vol. 46 No. 1, pp. 53–61

Hallikas, J., Karvonen, I., Pulkkinen, U., Virolainen, V.-M. and Tuominen, M. (2004), "Risk management processes in supplier networks", *International Journal of Production Economics*, Vol. 90, pp. 47–58

Latour, A. (2001), "Trial by fire: A blaze in Albuquerque sets off major crisis for cell-phone giants – Nokia handles supply shock with aplomb as Ericsson of Sweden gets burned – Was SISU the difference?", *Wall Street Journal*, January 29, A1

Svensson, G. (2000), "A conceptual framework for the analysis of vulnerability in supply chains", *International Journal of Physical Distribution & Logistics Management*, Vol. 30 No. 9, pp. 731–749

Tang, C. (2006), "Perspectives in supply chain risk management", *International Journal of Production Economics*, Vol. 103, pp. 451–488

Zsidisin, G., Ellram, L., Carter, J., and Cavinato, J. (2004), "An analysis of supply chain assessment techniques", *International Journal of Physical Distribution & Logistics Management*, Vol. 34 No. 5, pp. 397–413

Section I
Understanding and Assessing Risk in the Supply Chain

Chapter 2
Effective Management of Supply Chains: Risks and Performance

Bob Ritchie and Clare Brindley

2.1 Introduction

A significant feature of the rapidly evolving business climate spurred on by significant technology shifts, innovation, communication technologies and globalization, is the increasing prevalence of risk in almost every aspect of our lives. Whether real or imagined, we perceive greater exposure, increased likelihood and more severe consequences of already known risks whilst becoming aware of other risks previously unknown. FM Global (2007) concluded from their study of the views of 500 financial executives in Europe and America that most anticipated an increase in overall business risks in the foreseeable future. The top three risk areas featured global competition, supply chains and property-related risks. Individual organizations are continuously receiving information inputs identifying new risk sources, enhanced exposure to existing risks and escalating costs associated with compensation should such risks materialize. The emergence of risk management is an important response to such developments providing a contribution to most fields of management decision and control (e. g. Smallman, 1996; Giannakis et al., 2004). Supply Chain Risk Management (SCRM) represents the risk management response primarily to supply chain risks, although as will be seen later in the chapter, it has a much wider influence at the strategic enterprise risk level.

As the range, intensity and pace of developments in the 'risk management' field accelerates and organizations seek to mitigate their risk exposure, questions are being raised concerning the cost-effectiveness of risk management. Engagement in the processes and practices of risk management involves the commitment of resources and expenditure. The question posed is what is the impact of risk management on the performance of the organization? Against the explicit and implicit costs incurred by the organization is the need to match the impact on different performance criteria (e. g. profitability; security of supply; enhanced risk preparedness). In essence, what evidence is available to justify the SCRM investment in terms of benefits to corporate performance? The emerging field of Supply Chain Management is an appropriate field to evaluate such issues, since it has the capacity to demonstrate a diversity of risks together with risk management responses as well as producing an impact across most dimensions of an enterprise's performance. Brindley (2004) identifies global competition, technological change and the continuous search for competitive advantage as the primary motives behind organizations turning towards risk management approaches, emphasizing that the supply chains simultaneously represent the most important solution to such challenges whilst generating the most significant sources of risk. This view is re-affirmed by Christopher and Lee (2004) who recognize the increasing risk in organizational supply chains and identify the need for new responses to manage these.

The chapter is designed to explore the nature of the interaction between risk and performance, and on the basis of this improved knowledge and understanding to assess more effectively how the engagement in supply chain risk management strategies and activities might impact on corporate performance. The primary constructs of performance and risk are examined and the key linkages are identified, adopting an approach that addresses the perspectives of both researchers and practitioners and aiming to generate new and wider perspectives. The earlier sections review the significant developments in conceptual and empirical work, primarily in the fields of supply chain management, SCRM and corporate performance. These are not claimed to be exhaustive in relation to their particular field but rather sufficient to facilitate the linkages across all three fields. The nature and incidence of risk in the supply chain is explored in terms of risk and performance – the sources, drivers, consequences and management responses. The mutual interdependence of risk management and performance management are examined employing the context of the supply chain. The product of this development is the production of guidelines for practitioners and researchers which will support the generation of appropriate strategies to address particular risk drivers and the effective management of the risk management – performance interaction to maximize the benefits of SCRM.

2.2 Supply Chain Risk

2.2.1 Risk and Uncertainty

Agreeing a definition for the term *risk* has proved challenging for academics and practitioners alike for over a century, leading to the conclusion that there are probably as many definitions as writers on the theme (Ritchie and Marshall, 1992). The reason for this variety of definitions reflects different academic and professional disciplines and variations in the specific settings, decision contexts and problems being addressed. Sitkin and Pablo (1992, p. 9) reflect this in their generalized definition of risk as being *'the extent to which there is uncertainty about whether potentially significant and/or disappointing outcomes of decisions will be realized.'* Zsidisin (2003, p. 15) addressing the supply chain context more specifically defines risk as *'the potential occurrence of an incident or failure to seize opportunities with inbound supply in which its outcomes result in a financial loss for the [purchasing] firm.'* Most definitions (e. g. MacCrimmon and Wehrung, 1986) of risk comprise three common elements:

Likelihood of occurrence of a particular event or outcome,
Consequences of the particular event or outcome occurring,
Exposure or Causal pathway leading to the event.

Conceptually, these three elements appear to be readily identifiable and measurable, lending themselves to formulaic and precise resolution (e. g. rolling the dice at the casino). However, transferring these concepts to the practical business environment within risk management yields a totally different set of challenges. The *likelihood of occurrence*, more usually termed the probability, can be expressed in objective terms or in subjective terms, each being capable of measurement. Objective measurement relies on previous records of the occurrence of such events. Subjective assessment of the likelihood of occurrence relies more on the translation of previous experience and intuition. In practice there is likely to be the application of subjective judgments on any objective data.

Consequences are typically expressed as a multiple of simultaneous outcomes, many of which interact with one another (e. g. failure of a new product launch may generate consequences for the organization's reputation, financial performance and the standing of the individual product champion). Consequences should not simply be regarded as only or primarily negative, since the essence of risk taking is the potential opportunity to produce positive outcomes (Blume, 1971). The third element of the risk construct, the *causal pathway*, has particularly important implications for risk management. Understanding the nature, sources and causes of factors that generate the events or circumstances which might influence the type and scale of consequences (i. e. both positive and negative), and the likelihood of their occurrence are fundamental requirements for effective risk management.

Line A in Fig. 2.1 demonstrates a classic linear and consistent relationship through all levels of risk and outcomes considered (i. e. risk and performance out comes change directly in proportion with each other). Line B on the other hand

Fig. 2.1 Risk and performance relationships

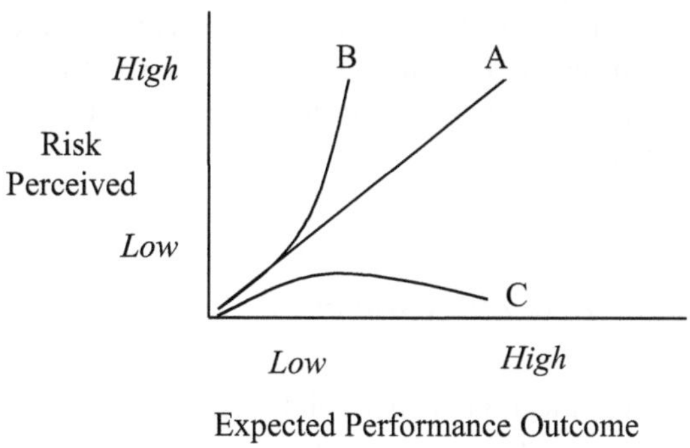

suggests some deterioration – as the risk perceived increases beyond a particular level then expected improvements in performance decline rapidly until the point is reached where improvements cease altogether and potentially become negative. For example, if the business continues to invest more resources into managing higher risks (e. g. buying information to refine risk sources and triggers, managerial time on data analysis, insurance cover for potential consequences) the cost of these actions may directly influence performance. Line C for completeness illustrates another commonly experienced risk-management-performance profile. As risk management activities are applied then the rate of perceived increase in risk slows reflecting the success of risk management, performance continues to increase and risk flattens off.

For our present purposes we consider that the essence of risk and risk taking involves the preparedness of the individual or organization to expose themselves to adverse outcomes on the basis that they may also achieve positive outcomes. The important point is recognizing that risk taking has 'upside' benefits as well as the 'downside' costs, although the emphasis is usually placed on the latter. Risk itself is not solely assessed in terms of the outcomes, positive or negative, but in terms of the scale of such outcomes and the likelihood of the outcomes happening. Thus risk may be portrayed as the acceptance of less desirable consequences or outcomes in return for the opportunity to achieve more desirable outcomes.

The terms risk and uncertainty are frequently used interchangeably in practice, although technically they describe distinctive situations. Uncertainty is viewed by many authors as a special case of the risk construct (Paulsson, 2004), in which there is insufficient information (Rowe, 1977), knowledge or understanding to enable the decision taker to identify all of the potential outcomes (Ritchie and Marshall, 1993), their consequences or likelihood of occurrence (MacCrimmon and Wehrung, 1986). Uncertainty typically relates to the situations where there is an absence of certain parameters such as potential outcomes, likelihood of each occurring and the consequences if they do. Pure risk sits at the other end of the continuum, typically representing the fully defined scenario of potential outcomes, objective probabilities of occurrence and fully determined consequences from each outcome. The further into the future the decision context, variables and outcomes under consideration (e. g. long-term strategic planning) the greater the difficulty in identifying, estimating and measuring these parameters, or in other words the greater the uncertainty involved. The terms risk and uncertainty are frequently used interchangeably, as typically risk contexts involving decisions are often somewhere in the middle of the risk-uncertainty spectrum (i. e. neither pure 'objective' risk taking nor complete uncertainty).

2.2.2 Risk and Outcomes

The clarification of four points relating to risk and outcomes may prove helpful at this stage, although we return to performance measures in more detail subsequently.

Firstly, the term *outcome* relates to an infinite range of performance measures and not just those associated with financial performance. Although much of the early research in the risk field related to financial decision contexts (e. g. Knight, 1921), more contemporary studies have argued for the inclusion of wider dimensions of performance criteria (e. g. Ritchie and Brindley, 2008) as risk management is increasingly applied in more diverse fields (e. g. Supply Chain Management). Secondly, the adverse consequences need not necessarily be negative or indeed significantly lower than the positive outcomes sought, providing that the degree of desirability of the positive outcomes is sufficiently attractive to accept the possible consequences of failure. In other words, if we are indifferent to the range of outcomes that may materialize then there is no reason to contemplate actions which seek to influence the nature or likelihood of these outcomes. Thirdly, risk is somewhat meaningless if divorced from those involved in taking the decision. Consideration of the nature of the impact on the individuals or the group involved represents an important dimension of risk appraisal. Personal perceptions of the likelihood and the potential impact will vary between individual group members and hence groups. In theory, the individual ought to be prepared to expend resources to reduce either the likelihood until the investment (marginal cost) is equivalent to the anticipated benefits (marginal revenues). Finally, the normal expectation in most decision situations is that risk is directly and inversely related to performance. Hence, an increase in risk as the determining variable is more likely to lead to higher outcomes in performance terms with the obverse being lower levels of risk resulting in lower performance outcomes. It is not strictly true to categorize risk as the independent variable and performance outcomes as the dependent variable. We cannot for example stipulate that higher risk taking causes better performance. More correctly we have an association between the two variables such that higher risk and higher performance are often equated with each other. In cases where the initial perception is one of low risk this does not imply poor performance since such profiles may represent the vast number of routine activities carried out in most businesses, providing the 'cash cow' for the business performance.

In a given situation, the initial risk perceived will be an assessment by the stakeholders or decision makers concerning the anticipated performance and the potential effects that risk may have on the performance outcomes. If there is a sufficient level of concern about the risks involved then the decision making unit may institute certain actions designed to obviate or mitigate such risks. Three broad categories of actions may be observed, including, searching for further information to clarify or resolve the risks; assessing the possible options to manage the risk sources, risk drivers and likely occurrence; and undertaking insurance cover and other management actions to mitigate against the consequences for performance should the risks materialise. There is in a sense an iterative loop with the feedback of information from the review of actions taken being followed by the revision of the risk perception and then perhaps a further iteration of the process (Fig. 2.2). Further iterations of this process will continue until either the decision makers feel that they have sufficiently resolved the risks or alternatively exhausted the available resources and timescales.

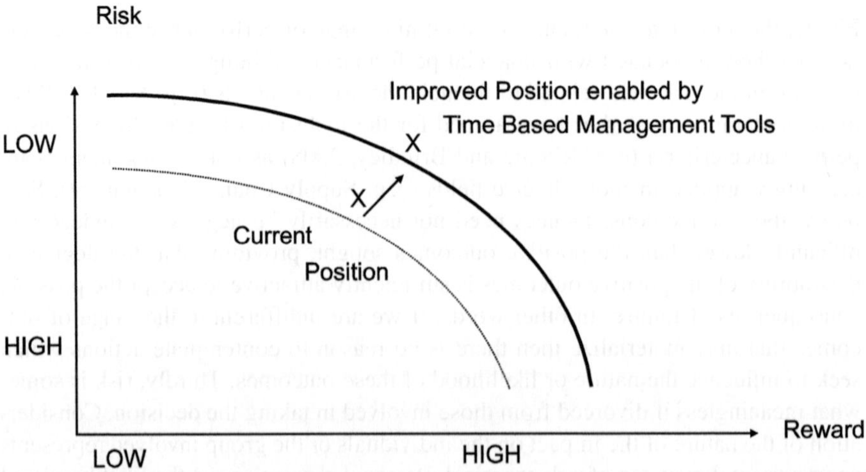

Fig. 2.2 Perfomance and risk assessment in profiling process

Risk then has a number of components which drive the perception:

Decision making unit and stakeholders,
Complexity and dynamism,
Level of aggregation,
Portfolio,
Time frame,
Decomposition of risk parameters,
Distribution and sharing.

The *stakeholders* and the members of the *decision making unit* need not be the same, indeed the shareholders in a business organization may represent the primary stakeholders yet not be directly engaged in the business risk management process. Correspondingly, managers involved in the decision making unit ought to address the interests of all of the stakeholder groups, recognizing that these may at times be inconsistent and at odds with each other. The shareholder may desire performance outcomes related to share price gains in the longer term whilst the commercial lender may seek more immediate assurances about short term profits, cash flows and solvency. Mixed in with such external interests are those of the individual members of the decision making unit who may gain from performance-related bonuses, employment security and status as a successful manager. Disentangling fully such interests and aspirations is seldom possible, although there has to be some clarity on the prioritization of these complementary and conflicting demands by the decision making unit to guide decisions within risk management. Added to this *complexity* is the *dynamism* or flux inherent in many decision situations, especially those experiencing above average risk levels. Indeed, the belief or perception that the situation under consideration may be subject to continuous change will in itself engender a greater perception of risk, especially if the scale of the changes and consequences are high.

The *level of aggregation* of the decision will often influence the perceived level of risk involved. If decisions are treated independently and in isolation from each other the business may remain quite comfortable with the individual risks involved. However, if these individual decisions are then aggregated or considered as a whole then the overall perception of risk is likely to change, most probably in an upward direction (i. e. more risky). It is important for the organization to determine the appropriate level in the hierarchy for the aggregation of the risk profile. Associated with the level of aggregation is the concept of a *Portfolio* of risk and performance outcomes. The consideration of a business as comprising a number of investments or projects at any one time represents an important feature associated with risky decisions and their impact on performance. This portfolio of investments approach enables a group of activities, investments or projects to be viewed together rather than in isolation which means that individual projects or investments may provide offsetting contributions to performance and risk parameters. Early work in the financial economics field dealing with investment risk and returns resulted in a number of theoretical developments (e. g. Capital Asset Pricing Method – Ball and Brown 1968) which have resulted in some interesting and useful guidance for practical decisions in securities markets but have failed to provide similar benefits in the 'messier, partial and fragmented' situation found in most organizations.

The *timeframe* is an important consideration as there is an evident relationship between the perspective of time and the perception of risk. Focus on the immediate or short term risks may result in decisions that effectively resolve the present position but expose the business to higher risk exposure in the longer term. The converse may equally apply when focusing on the longer term and ignoring the short term consequences of the decisions taken. Another notable feature relating to the timeframe concerns the anticipated length of time from the decision being taken to the realization of the outcomes or consequences. It is likely that the shorter the anticipated elapsed period from the decision to outcome realization, then the higher the predictability or greater confidence there will be and hence lower perceptions of risk (i. e. short term decisions may on balance be perceived as less risky than long term decisions).

Faced with a risky decision there is a natural tendency to explore the nature and scale of the risks perceived. In essence this results in the *decomposition of risk parameters.* The decision maker may seek to divide the situation encountered into a number of component parts and to seek more information about each in the anticipation that this process of investigation will aid in gaining a better understanding of the relationships involved thereby resolving even if not reducing the risks perceived.

A further important feature associated with risk and its management relates to the *distribution and sharing* of risks. There has been a long-standing tradition in the financial services sector (e. g. banks and insurance businesses) of distributing large risks amongst a number of businesses such that the exposure for any single institution is kept low, on the basis that the realization of the risk will have less than disastrous consequences for any single business. The supply chain context provides a similar opportunity to share or distribute certain risks across the members of the chain. This may be far more achievable for the operational level decisions (e.g quality assurance, avoidance of stock-outs) than the more strategic level though the development of alliances with supply chain partners may provide risk sharing opportunities for strategic developments into new markets.

2.3 Supply Chain Risk Management

There may not yet be agreement on the definition of Supply Chain Risk Management (SCRM) but there is agreement on the main components of SCRM, although these may be differently termed. Most definitions would incorporate the following clusters of activities:

- *Risk identification and modeling* – including the sources and characteristics of risks, what may trigger them and the relationship to the supply chain performance in terms of effectiveness and efficiency.
- *Risk Analysis, Assessment and Impact Measurement* assessing the likelihood of occurrence and potential consequences.
- *Risk Management* – generating and considering alternative scenarios and solutions, judging their respective merits, selecting solutions and undertaking the implementation.
- *Risk Monitoring and Evaluation* – monitoring, controlling and managing solutions and assessing their impact on business performance outcomes.
- *Organizational and Personal Learning including Knowledge Transfer* – seeking to capture, extract, distill and disseminate lessons and experiences to others within the organization and its associated supply chain members.

These components of the SCRM approach represent an integrated decision making approach and one which interacts extensively with other members of the supply chain. In essence, SCRM represents a more pro-active approach to managing risks and performance in the supply chain in advance to avoid or minimize potential undesirable consequences. Such a pro-active approach does not necessarily ensure that all such potential risks can be identified in advance or if identified, sufficiently well resolved to prevent some or all of the undesirable consequences. SCRM like other management approaches is dependent on good quality management in terms of knowledge, abilities, experiences and skills. The concepts, tools and technologies provide support but are unable to replace the judgments required in most risky decision situations. SCRM of necessity covers a number of levels of activities, from operational through tactical to strategic since in many supply chains these are inseparable. For example, procedures for monitoring in-bound component quality provide an early warning signal of potential disruption or complaints.

SCRM in common with other organizational functions, shares a number of more generic characteristics and influences associated with the corporate level of organizational decision making involving risk (Ritchie and Marshall, 1993). These include the nature of the relationship between risk and performance at the corporate level;

the distribution of risk across the existing and prospective portfolio of corporate investments; the capacity to redistribute or share risks internally and externally; the time-span over which risks and outcomes are normally measured and assessed; and the attitude or the preparedness of the organization to engage with risk and its consequences. These factors will be present in most business organizations although may vary depending on sector, history, prior experiences, decision makers and other contextual factors. It is true that much of the early work in these elements has related to the financial and securities markets although increasingly attention has been directed to application in other business sectors. The remainder of the discussion in the present chapter will focus on supply chain risks and their management. Several authors (e. g. Child and Faulkner, 1998) have argued that decisions relating to changes in the supply chain structure and relationships ought to involve the analysis and evaluation of the associated potential outcomes in terms of benefits, costs and risks. Lonsdale and Cox (1998) likewise contend that performance and risk are inextricably interconnected and thus require the development and application of supplier management tools and controls to maximize performance whilst controlling the consequential risks.

2.4 Performance

The second of our key terms, *performance,* has also generated an almost infinite variety of definitions. Even within the general confines of business performance there will be variations relating to specific sectors, contexts or functional perspectives. Anthony (1965) provided a generic and now well-established definition of performance, dividing this construct into two primary components, efficiency and effectiveness. Efficiency addresses performance from a resource utilization perspective indicating that greater efficiency is derived from producing a greater quantity of outputs (i. e. products and/or services) for a given volume of resource inputs. Effectiveness addresses performance related to the degree to which the planned outcomes are achieved (e. g. achieving the objective of avoiding supply disruptions during a given period may be viewed as an effective outcome). Colloquially, effectiveness is often described as ‘doing the right things’ whilst efficiency may be expressed as ‘doing things well.’ However, these two primary components need not necessarily be consistent and almost certainly do not demonstrate the practice of operating in unison. For example, the avoidance of supply disruptions during the period may have required maintaining high buffer stocks or special incentive payments to the supplier. Such actions may well prove highly effective against a requirement to enhance customer service and satisfaction levels, but prove highly inefficient in terms of profitability. Earlier definitions of aggregate business performance tended to focus on the efficiency dimension, featuring financial performance and other quantifiable parameters as the primary outcome measure. Subsequently, more encompassing definitions of performance have evolved, most notably the Balanced Scorecard (Kaplan and Norton, 1992, 1996) which incorporates not only the Financial Perspective but also the Internal Perspective, the Customer Perspective and the Innovative and Learning Perspective. The search continues to determine appropriate performance measures and metrics which can be adopted or adapted to all fields of business activity, providing easily captured and communicated; readily understood; reliable and robust measures to support the business unit in terms of effectiveness and efficiency. Melnyk et al. (2004, p. 210), reflecting on metrics and performance measurement concluded that 'performance measurement continues to present a challenge to operations managers as well as researchers of operations management'.

2.4.1 Performance Criteria

There are several performance criteria associated with most supply chains, although their relative importance may change depending on the specific supply chain context, timeframe, relationships etc. These may vary in terms of significance depending on the position in the chain itself and the position from which it is viewed. These criteria may also be more or less tangible and hence more or less difficult to measure. The more tangible measures include:

Timeliness – the availability of the components or services at the agreed time, recognizing the problems resulting from delayed delivery and often the storage, deterioration and stock management costs associated with earlier than agreed delivery.

Security of Supply – ensuring that complete interruptions or disruptions to the supply of goods or services are avoided.

Quality – is itself a multi-faceted dimension although principally the requirement is that the component or service is 'fit for purpose' as agreed in the specification between the supplier and customer at the contractual stage. The consequences of poor quality include downtime, reworking and subsequent failure to satisfy the customer's requirements in terms of quality and timeliness. Poor quality products or service, either upstream or downstream, may impact on the level of satisfaction of the customer with consequences for future revenues and possibly more immediate claims for financial compensation.

Price – although price may be fixed in many supply arrangements at the time of contracting there are many supply chain arrangements where price may be subject to variation (e. g. changes in local taxation, exchange rate movements). Volatility in terms of price may result in difficulties in passing on price changes to the customer and potentially have consequences in lost profit.

Associated Costs – the procurement, logistics and stock management costs may be largely predictable for those transactions which flow through the chain without any problems. However, difficulties arising from problems associated with timeliness, quality and price may absorb significant time and costs to remedy.

Support – the level and quality of the support provided by suppliers in the chain may have a considerable impact in resolving adverse performance outcomes in terms of timeliness, quality and price and may indeed help to anticipate these.

Less tangible measures of performance, although no less important are the effectiveness of the communication channels; the willingness to share information with other supply chain members; the development of professional relationships; and ultimately the establishment of trust between the organizations and their members. Other factors such as the reputation of the firm, often generated by issues not directly related to the supply chain itself, may pose risks. Inadvertent comments by senior executives or the failure to endorse certain protocols may damage the reputation of the organization, with subsequent impact on the overall performance.

2.4.2 Risk and Performance – Timeframe

Time represents an important dimension when considering risk and performance. Risk perception is influenced by the passage of time. Perceptions at an early stage are likely to be influenced by high levels of uncertainty, largely reflecting a lack of or imprecise information and knowledge (e. g. what are the possible risks and key drivers?). The Risk Management process involving the investment of time, information gathering and data analysis may help to resolve the risks perceived, although not necessarily removing them. There is a corollary to this anticipated outcome in that gaining more information and knowledge about possible risks and outcomes may equally diminish confidence and heighten risk perceptions as new unforeseen risks are revealed. In 'one-off' decision we can envisage that the progression of time combined with the investment in risk management may ideally lead to a reduction in risk perception and increased confidence. In the continuous flow context associated with many supply chain activities and processes, the passage of time either without any incidents or incidents that have been well handled and foreseen may again build the confidence level of the risk management team. The difference in this latter context is that responses have to be made to risk outcomes as they occur including making changes immediately to avoid their recurrence in the continuous flow in the next cycle. Time may be an implicit parameter in the risk management process but is nevertheless important. The timeframe may also be used to categorize decisions, for example, operational, tactical and strategic. Paulsson (2004) utilizes this approach to differentiating supply chain risks into Operational Disturbance, Tactical Disruption and Strategic Uncertainty. Kleindorfer and Wassenhove (2003), however, subdivided supply chain risks into only two categories Supply Co-ordination Risks and Supply Disruption Risks, dividing Paulsson's (2004) tactical level decisions between the two categories. Whilst such categorization may prove helpful conceptually, Mintzberg and Waters (1985) highlighted the difficulty in differentiating these in practice, suggesting that strategic decisions are in essence the aggregation of a sequence of operational and tactical decisions, leading to some common planned or emergent pattern. Similarly, the differentiation between tactical and operational decisions may prove somewhat arbitrary in practice.

2.5 Performance and Risk Metrics

The inter-relationship between the investment in SCRM and the outcomes in terms of changes in performance and risk (i. e. the resulting risk and performance profiles for the organization as a whole) are increasingly important aspects of SCRM. The measurement of corporate performance and risk may be addressed at different levels of aggregation and from a number of different although not mutually exclusive perspectives. Different stakeholders, both external and internal, may seek a different balance in terms of the timeframe (i. e. short-term versus long-term performance), emphasise different criteria (e. g. profitability, cash flow or customer service) and display different attitudes to risk exposure (e. g. avoidance of excessive risk). Stainer and Stainer (1998) identified eight different categories of stakeholder (i. e. shareholders, suppliers, creditors, employees, customers, competitors, Government and society) together with their differing performance expectations from the business. A common feature of most, if not all, stakeholders is the underlying requirement for generating performance in terms of profit consistent with the risks associated with achieving this performance. Mathur and Kenyon (1997) support the importance of financial performance metrics as providing the basis for confidence in the business providing reassurance to all stakeholders including those associated with the wider societal concerns (e. g. Stainer and Stainer, 1998), There is no question that financial performance outcomes remain the primary metrics for commercial situations such as SCRM. Within this there may be different emphases, focusing on more long-term perspectives rather than short term timeframes often implied by many financial performance measures (Marsden, 1997).

A generic model of corporate performance incorporating the risk measure links performance to the three independent variables Industry Characteristics (IC), Strategic Decisions (S) and Risk (R):

$$\text{Performance} = f(\text{IC}, \text{S}, \text{R}) .$$

Risk in turn is viewed as being determined by the two variables Industry Characteristics and the Strategy developed. As Ritchie and Marshall (1993, p. 165) concluded 'a well devised strategy could simultaneously reduce risks and increase returns'. The generic performance metrics typically employ a financial definition of performance (e. g. return on assets (ROA); return on investment (ROI)) and generally seek to maximize this performance over the longer term period as opposed to the short term such as the current year.

The SCRM Framework (Fig. 2.3) comprises two major groups of components. The more strategic and enduring group of components are located at the core–risk profile, performance profile, strategic timeframe and the stakeholders participating in the supply chain. The outer ring comprises the key components or activities involved in the risk and performance management process.

Considering initially the core components:

1. The *Performance Profile* represents a multi-dimensional view of the organization (i. e. equivalent say to the Balanced Scorecard (Kaplan and Norton, 1992, 1996) incorporating Financial, Internal, Customer, Innovative and Learning perspectives. This profile engenders the strategic aims of the organization and its stakeholders, recognizing the need for flexibility and balancing between the various, often incompatible, performance criteria and stakeholder demands.
2. Similarly, the *Risk Profile* represents the scale and nature of the risk exposure that the organization is prepared to accept. In many respects this profile comprises a portfolio of different types of risk and the organization may well be prepared to accept high risk exposure in one element of the business if this is counter-balanced by lower risk exposure elsewhere in the portfolio of projects/activities/investments.
3. The *Timeframe*, as discussed earlier, indicates the importance of locating the point in time at which risk and performance are viewed, assessed and managed.
4. The inclusion of the *Supply Chain Stakeholders* as a component in the risk/performance core is designed to emphasize the inter-dependence of the members of the supply chain. The performance and risk profiles for the organization are dependent to differing degrees on the other partners and stakeholders in the supply chain.

The outer ring (Fig. 2.3) represents the ongoing processes of risk and performance management. The Risk Management components comprise a sequence of

- identifying risk sources and drivers,
- assessing their potential impact,
- instituting appropriate risk management responses,

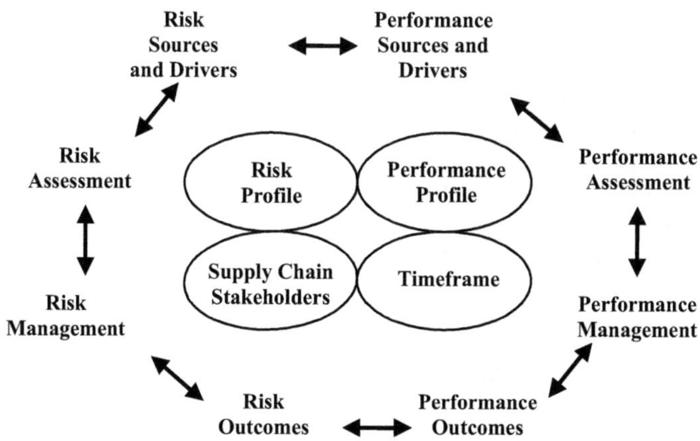

Fig. 2.3 Supply Chain Risk Management Framework – Process overview

- evaluating the impact on risk outcomes.

The performance management sequence parallels that for risk management.

- identifying performance sources and drivers,
- assessing their potential impact,
- instituting appropriate performance management responses,
- evaluating the impact on performance outcomes.

Three important points relate to the interpretation of the risk and performance management process elements in the Framework:

1. The sequence of actions within each of the processes is not linear nor necessarily progressive from one to the other, it often requires the re-iteration of previous stages to achieve greater understanding or verification.
2. The risk and performance processes are inextricably joined in practice with decision makers continuously involved in balancing risk and performance information.
3. The outer ring of more tactical or operational activities involved are informed by and also inform the core components, namely the performance profile, risk profile and stakeholder profile, albeit often from a different timeframe.

2.6 Risk-Performance: Sources, Profiles and Drivers

Risk and performance in the supply chain context are influenced by a variety of different factors. Structuring these factors may prove helpful in appreciating the range and complexity of the situation faced in SCRM. A generic categorization based on the model initially developed by Ritchie and Marshall (1993) divides the sources and risk drivers in to seven groups of elements:

$$\text{Risk (R)} = f(E_r\ I_r\ SC_r\ SS_r\ O_r\ P_r\ DM_r)\,,$$

where:

E_r = environmental variables,
I_r = industry variables,
SC_r = supply chain configuration,
SS_r = supply chain stakeholders,
O_r = organizational strategy variables,
p_r = problem specific variables,
DM_r = decision-maker related variables.

Note: The Industry Characteristic (IC) variable referred to earlier in the Performance–Risk relationship model has been decomposed into the first three groups of element in the current model. The subscript ‘r’ seeks to emphasize the focus on the risk dimensions of each group of elements, especially those associated with risk sources and risk drivers.

These elements are translated into the upper boxes in Fig. 2.4. The elements have been further divided on the basis of those that are largely unavoidable or not capable of being influenced by the organization (i. e. *systematic risks)* on the left arm in the figure and those that are avoidable by the organization and result directly from its decisions and actions (i. e. *unsystematic risks)* on the right arm in the figure. The merging of the two arms indicates that all seven elements interact with each other as part of the SCRM process and that the management process requires the handling of both the unavoidable and avoidable risks generated. All of these elements relate to both risk and performance and either singly or in combination, determine the risk and performance profile for the organization at that point in time and for that particular decision or set of decisions. This is a dynamic situation and any of these seven sources may generate new risks at any time on a continuous basis, affecting both the risk and performance profiles of the organization.

The sources identified in Fig. 2.4 are only representative of an almost infinite number of factors exposing the business to undesirable consequences in terms of performance and risk. The organization needs to establish which are critical and which are less so, with the former likely to be a very small proportion of the total. The term *driver* has been introduced to differentiate those factors most likely to have

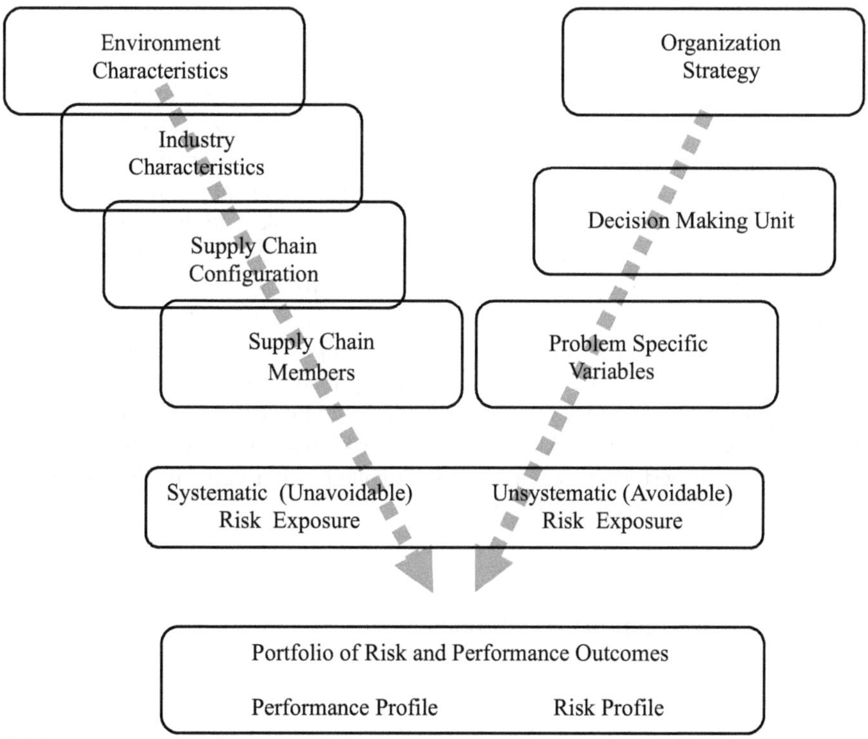

Fig. 2.4 Risk and performance: Sources and drivers

a significant impact on the exposure (i. e. sources, causal pathways, likelihood and consequences) to undesirable performance and risk outcomes. Performance drivers may well offer opportunities to enhance performance outcomes, albeit by incurring increased risk. For example the decision to develop a new direct channel to the consumer, bypassing existing distribution channel members would expose the business to new risks both from the reaction of the consumer and the retaliatory actions of the other channel members, although possibly improving potential performance outcomes. Key risk and performance drivers will usually be associated with each of the seven sources listed in Fig. 2.4. The dynamism referred to earlier suggests that the composition of key drivers may well vary over time and change as the supply chain situation changes in terms of structure and membership.

The elements in Fig. 2.4 relate to risks that pose a major threat to the survival and future development as well as those influencing the more 'normal' performance of the business in terms of effectiveness and efficiency. Management attention should focus on those drivers which are likely to have a significant impact on the performance or risk profile. Since SCRM concerns the interrelationships within the typical supply chain network, all members within a network will be potentially exposed to the risks, although the direct impact may be ameliorated or modified by the actions taken by others in the chain. Collectively, members throughout the supply chain may benefit if everyone engages in systematic SCRM activities.

The underlying presumption of SCRM is that the risk and performance sources and drivers can be foreseen, identified, evaluated, prioritized and managed. The practical reality suggests considerably more 'fuzziness' relating to these drivers, their impact on performance and the effectiveness of possible management solutions. This begs the question as to why organizations bother to prepare themselves for risk encounters and invoke SCRM. Quite simply, organizations believe that the best approach is to accept that they will be exposed to risks and the best strategy is to become more aware and pro-active towards the risks and better prepared to respond more quickly should such risks materialize (Kovoor-Misra et al., 2000). The simultaneous presence of different drivers may have a compounding effect both on the exposure and potential consequences. For example, the dislocation of supplies due to transport failures may be compounded by inadequate communications between supply chain members and further exacerbated by poor management controls in the principal organization. Alternatively, the risk drivers may be counterbalanced by high levels of performance, enabling the management to effectively control the worst consequences of supply disruption.

2.7 Risk Management Responses

The range of risk management responses is typically extensive, although some exemplars of risk management responses (Fig. 2.5) include risk insurance, information sharing, relationship development, agreed performance standards, regular joint reviews, joint training and development programmes, joint pro-active assessment and

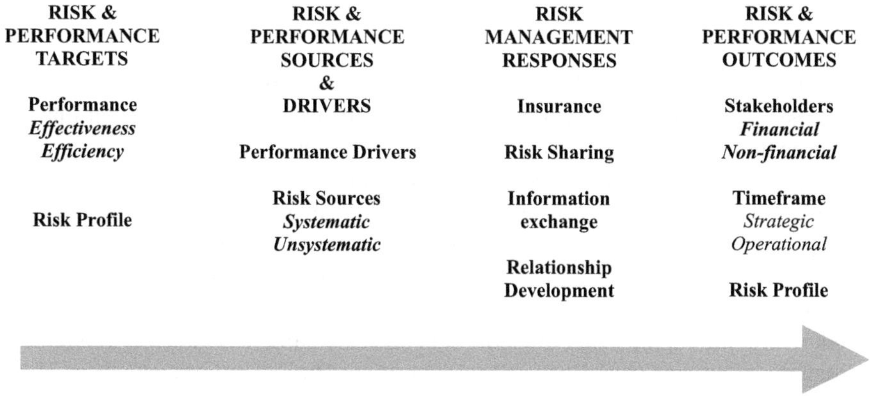

Fig. 2.5 SCRM framework

planning exercises, developing risk management awareness and skills, joint strategies, partnership structures and relationship marketing initiatives. A number of supply chain groups have displayed some degree of progression in the risk management responses, leading from the more individualistic and independent responses (e. g. insurance, establishing supplier service levels) to the more co-operative responses (e. g. sharing strategic information, relationship development and partnering). Similar developmental trends were discovered by Kleindorfer and Saad (2005) when examining disruption risks in supply chains in the US chemical Industry. Progressing to more integrated and comprehensive approaches may be the result of the nature and severity of the potential risk consequences, the number of members involved in the supply chain and enhancements in terms of confidence and trust between the parties engaged in the supply chain. The initiation of such developments is often undertaken by the larger organizations in the supply chain, since they may have the resource base and expertise, as well as the comparative power to encourage participation by smaller enterprises.

Risk management needs to target both the systematic and unsystematic risk sources. Whilst the organization may not be able to control sources of systematic risk (e. g. changes in interest rates or political instability), the adoption of particular management approaches may enable the organization to better understand the risks and to respond more quickly and effectively to modify or ameliorate the impact on performance (e. g. reducing financial risk exposure in advance by limiting borrowing). Similarly, improving the awareness and understanding of unsystematic risks can enable the development of strategies which either avoid or minimize exposure to identified risks. Equally, developing relationships and partnerships with key members of the supply chain may provide a more generic shield against exposure to other unsystematic risks emanating from elsewhere in the supply chain.

2.8 Conclusions

The chapter aimed to develop an understanding of the main constructs underpinning risk and performance within the supply chain, especially the inherent linkage between these two. The dimension of time was introduced as another key dimension influencing risk perceptions (i. e. uncertainty reduction and risk resolution as time progresses. Risk Management has been implicit in the management of supply chains over a long period of time. More recently (see Mentzer, 2001; Brindley, 2004) a concerted effort has been made to study these in a more logical and coherent manner. This was in response to the significant pressures experienced both contextually and from within the supply chain to modify and in some situations radically change the structures, modes of operations and relationships. These developments themselves may further engender an increased sense of uncertainty and risk throughout the various stages of the supply chain.

The evidence suggests that organizations are increasingly recognizing the interaction between performance and risk. There is also a trend towards risk and performance metrics which are much more widely based (i. e. not dependent primarily on quantitative, financially oriented and short-term metrics). The new metrics are evolving and are designed to tackle some of the broader risk-performance issues at the enterprise level (i. e. Enterprise Risk Management systems) and are more strategically oriented. The development of improved relationships leading to more formalized associations is an evident trend in many supply chain situations with the emergence of issues such as confidence and trust emerging in the discussions and developments between supply chain partners. The recognition that SCRM can create both positive and negative consequences for the businesses in the supply chain highlights the importance of developing reliable, robust and practical measurement systems. Such multi-dimensional risk-performance measures need to be well understood and applied appropriately. The chapter has sought to make a contribution to this requirement.

References

Ambler, T. (2000), Marketing and the bottom line, Pearson Education, UK

Anthony, R.N. (1965), Planning and Control Systems: A Framework for Analysis, Harvard University School of Business Administration, Division of Research, Boston

Baker, M.J. (1976), "The written analysis of cases", Quarterly Review of Marketing, Vol. 1, pp. 1–6

Ball, R. and Brown, P. (1968), "An empirical evaluation of accounting income numbers", Journal of Accounting Research, Vol. 6, No. 2, pp. 159–178

Bettis, R.A. (1982), "Risk considerations in modelling corporate strategy", Academy of Management Proceedings, pp. 22–25

Bettis, R.A. and Hall, W.K. (1982), "Diversification strategy, accounting determined risk and accounting determined return", Academy of Management Journal, Vol. 25, No.2, pp. 254–264

Bowman, E.H. (1980), "A risk/return paradox for strategic management," Sloan Management Review, Vol. 21, No. 3, pp. 17–31

Blume, M.E. (1971), "On the assessment of risk", Journal of Finance, Vol. 26, No. 1, pp. 1–10
Brindley, C.S. (ed.) (2004), Supply Chain Risk, Ashgate Publishing Ltd., UK
Brindley, C.S. and Ritchie, R.L. (2001), "The information-risk conundrum", Marketing Intelligence and Planning, Vol. 19, No. 1, pp. 29–37
Burtonshaw-Gunn, S.A. and Ritchie, R.L. (2004), "Developments in construction supply chain management and prime contracting", Proceedings of the 1st International SCRI Symposium, 30th–31st March 2004, pp. 334–347, Blackwell, UK
Christopher, M. and Lee, H. (2004), "Mitigating supply chain risk through improved confidence", International Journal of Physical Distribution and Logistics Management, Vol. 35, No. 4, pp. 388–396
Doherty, N.A. (2000), Integrated Risk Management: Techniques and Strategies for Reducing Risk, McGraw-Hill, New York
Eisenhardt, K.M. (1989a), Agency Theory: An Assessment and Review, Academy of Management Review, 14(1), 57–74
Eisenhardt, K.M. (1989b), "Building theories from case study research", Academy of Management Review, Vol. 14, No. 4, pp. 532–550
Ellram, L.M. (1996), "The use of case study method in logistics research." Journal of Business Logistics, Vol. 17, No. 2, pp. 93–128
FM Global (2007), "Managing Business Risk – Through 2009 and Beyond", FM Insurance Company Limited, Windsor, Berks., UK
Giannakis, M., Croom, S., and Slack, N. (2004), "Supply chain paradigms", in: Understanding Supply Chains, New, S. and Westbrook, R. (eds.), Oxford University Press, UK, pp. 1–22
Holton, G.A. (2003), Value-at-Risk, Academic Press, USA
Kaplan, R.S., Norton, D. (1992), "The balanced scorecard measures that drive performance", Harvard Business Review, January/February, 71–79
Kaplan, R.S., Norton, D. (1996), "The balanced scorecard", Harvard Business School Press, Boston, MA, USA
Kleindorfer, P.R. and Saad, G.H., (2005), "Managing disruption risks in supply chains", Production and Operations Management, Vol. 14, No. 1, pp. 53–68
Kleindorfer, P.R. and Wassenhove, L.K. (2003) "Managing Risk in Global Supply Chains", Paper presented to Wharton Insurance and Risk Management Department Seminar, 27th February 2003, Wharton University
Knight, F.H. (1921), "Risk, Uncertainty and Profit", Houghton Mifflin Company, Boston and New York
Kovoor-Misra, S., Zammato, R., and Mitroff, I.I. (2000), "Crisis preparation in organisations: Prescription versus reality", Technological Forecasting and Social Change, Vol. 63, pp. 43–62
Latham, Sir M. (1994), Construction the Team: Joint Review of Procurement and Contractual Arrangements in the United Kingdom Construction Industry Final Report, HMSO, London
Lonsdale, C. and Cox, A. (1998), Outsourcing: A Business Guide to Risk Management Tools and Techniques, Eastgate Press, Boston, Lincs
MacCrimmon, K.R. and Wehrung, D.A. (1986), Taking Risks: The Management of Uncertainty, Free Press, New York
Mallman, C. (1996), "Risk and organizational behaviour: A research model", Disaster Prevention and Management, Vol. 5, No. 2, pp. 12–26.
Markowitz, H.M. (1952), "Portfolio selection", Journal of Finance, Vol. 7, No. 1, pp. 77–91
Marsden, C. (1997), "Corporate citizenship", Faith in Business, Vol.1, No. 4, pp. 3–15
Mathur, S.S. and Kenyon, A. (1997), Creating Value – Shaping Tomorrow"s Business, Oxford, Butterworth-Heinemann Ltd., pp. 25–43
Melnyk, S.A., Stewart, D.M., and Swink, M. (2004), Metrics and performance measurement in operations management: Dealing with the metrics maze, Journal of Operations Management, Vol. 22, pp. 209–217
Mentzer, J.T. (ed.) (2001), Supply Chain Management, Sage Publications Ltd., USA

Miles, M.B. and Huberman, A.M. (1994), Qualitative Data Analysis: An Expanded Sourcebook, 2nd Edn., Sage Publications, London

Mintzberg, H. and Waters, J.A. (1985), "Of strategies, deliberate and emergent", Strategic Management Journal, Vol. 6, pp. 257–272

Mintzberg, H., Ahlstrand, B., and Lampel, J. (1998), Strategy Safari, Free Press, New York

Mitchell, J.C. (1983), "Case and situation analysis", The Sociological Review 31, cited in Smith and Dainty (eds.) (1991), The Management Research Handbook, Routledge, London

Newall, J. (1977), "Industrial buyer behaviour", European Journal of Marketing, Vol. 11, No. 3, pp. 166–211

Paulsson, U. (2004), "Supply chain risk management" in Brindley C. (ed.), Supply Chain Risk Management, Ashgate, UK, pp. 79–96

Peck H. (2005), "Drivers of supply chain vulnerability: An integrated framework", International Journal of Physical Distribution and Logistics Management, Vol. 35, No. 4, pp. 210–232

Porter, M.E. (1996), "What is Stratgey?", Harvard Business Review, November–December, pp. 61–78

Ritchie, R.L. and Brindley, C.S. (2004), "Risk characteristics of the supply chain – a contingency framework", in Brindley C.S. (ed.) Supply Chain Risk, Ashgate Publishing, UK

Ritchie, R.L. and Brindley, C.S. (2000), "Disintermediation, Disintegration and Risks in the SME Global Supply Chain", Management Decision, Vol. 38, No. 8, pp. 575–583

Ritchie, R.L. and Marshall, D.V. (1993), Business Risk Management, Chapman Hall, London

Rowe, W.D. (1977), Anatomy of Risk, Wiley, New York

Scappens, R.W. (1990). "Researching management accounting practices: The role of case study methods", British Accounting Review, Vol. 22, pp. 259–281

Shepherd, C. and Gunter, H. (2006), "Measuring supply chain performance: Current research and future directions", International Journal of Productivity and Performance Management, Vol. 55, No. 3/4, pp. 242–258

Sitkin, S.B. and Pablo, A.L. (1992), "Reconceptualizing the determinants of risk behaviour", Academy of Management Review, Vol. 17, No. 1., pp. 9–38

Smith, N.C. and Dainty, P. (eds.) (1991), The management research handbook, Routldege, London

Stainer, A. and Stainer, L. (1998), "Business performance – A stakeholder approach", International Journal of Business Performance Management, Vol.1, No. 1, pp. 2–12

Valla, J.P. (1982), "The concept of risk in industrial buyer behaviour", Presented at workshop on organisational buyer behaviour, European Institute for Advanced Studies in Management, Brussels, December 9–10

Warren, F. (1992), Introduction in: Royal Society Study Group, Risk Analysis, Perception and Management, The Royal Society, London

Wright, P., Pringle, C., and Kroll, M. (1992), Strategic Management Text and Cases, Allyn and Bacon, Needham Heights, MA

Yin, R.K. (1994), Case Study Research: Design and Methods, Sage, Newbury Park, CA

Zsidisin GA (2003), Managerial Perceptions of Supply Risk, The Journal of Supply Chain Management. A Global Review of Purchasing and Supply Vol. 39, No. 1. pp. 14–25

Chapter 3
Managing Supply Chain Disruptions via Time-Based Risk Management

ManMohan S. Sodhi and Christopher S. Tang

Abstract We wish to motivate research on the practice of preplanned response to rare events that can disrupt the supply chain. We present a time-based risk mitigation concept and illustrate how this concept could enable companies to reduce the impact of such events while potentially increasing their competitiveness. The underlying idea is that if a firm can shorten the response time by deploying a recovery plan soon after a disruption, then this firm can reduce the impact of the disruption by way of fast recovery. We present examples from a wide variety of contexts to highlight the value of time-based risk management.

Key words: Time-based competition; supply chain disruption; supply chain risk management; business continuity; contingency planning; supply chain responsiveness

3.1 Introduction

Our aim is to motivate research on the practice of systematic and preplanned response to rare events that can cause huge disruption to the supply chain. Such events are costly to prevent and companies may be reluctant to invest in prevention as the returns are unclear. However, they may be able to respond effectively to such risk incidents after they occur by containing their impact through quick response. We provide a simple framework to think about response and motivate further research and improved practice through a variety of examples both within and outside supply chain management.

Our proposal extends *business continuity* efforts in practice from a local context to a supply-chain wide one. We break up response to supply chain disruptions into three time elements – *detect* the event across the supply chain ($D1$), *design* a response ($D2$), and *deploy* the response ($D3$). We refer to these elements as the 3-D framework. By focusing its efforts on ensuring that systems and processes are in place to reduce to these three time elements, a company reduces the overall *response* time ($R1$) and thus *recovery* time ($R2$) and total impact. We illustrate this concept through examples from a variety of contexts.

Our contribution is to highlight a potentially rich area of empirical and modeling research that can complement the existing literature that focuses on prevention rather than post-incident recovery. Time-based risk management dovetails into the company's business continuity efforts and provides a basis for risk reporting for its lenders and investors. With time-based risk management, investment in risk management is low while increasing competitiveness due to improved responsiveness through more awareness of supply chain processes and of disruptions as well as more communication across the supply chain within the company and with its partners.

This chapter presents the concept of time-based risk management, its importance through a variety of examples, and its implementation. We provide avenues for research and the basis for improved practice.

3.2 Background

Many companies are expanding their supply chains to more external partners in different countries as a way to reduce cost and product development cycle and to explore new markets. For example, Boeing increased the outsourced content from 50% to 70% when developing its new 787 model and spread its suppliers over 20 countries. Also, according to an industry study conducted by AMR in 2006, over 42% of the companies manage more than 5 different global supply chains for different products in different markets. As supply chains become more complex, companies find their supply chains more vulnerable to unforeseen disruptions – rare but severe events that disrupt the flow of goods and information in a supply chain. Without a disruption management system in place, these disruptions can have huge impact in terms of cost and recovery time to the company and its customers.

Examples of impact include Ericsson losing 400 million euros in the quarter following a minor fire at their supplier's semiconductor plant in 2000. In addition, due to a design flaw of the Pentium microprocessors, the recall of 5.3 million chips has cost Intel $ 500 million in 1994. Furthermore, New Orleans has not fully recovered as of 2008 – three years after the landfall of Hurricane Katrina in August 2005. Over 100 patients have died in 2008 as a result of blood thinning drug Heparin contaminated with unsafe substance (Pyke and Tang, 2008).

Based on an analysis of 827 disruption announcements made over a 10-year period, Hendricks and Singhal (2005) found that companies suffering from supply chain disruptions experienced 33–40% lower stock returns relative to their industry benchmarks over a 3-year time period that starts one year before and ends two years after the disruption announcement date.

Other examples of significant supply chain disruption include Mattel's recall of over 18 millions of toys in 2007 (Casey and Pasztor, 2007). Dell recalled 4 million laptop computer batteries made by Sony in 2006. Land Rover laid 1400 workers off after their supplier became insolvent in 2001 as production could not continue without parts. Dole suffered a large revenue decline after their banana plantations were destroyed after Hurricane Mitch hit South America in 1998. Among the many instances of disruptions after 9/11 attacks in 2001, Ford had to close five plants for several days owing to the suspension of all air traffic. For more details, see Martha and Subbakrishna (2002), and Chopra and Sodhi (2004).

Supply-chain disruptions are getting CEOs' attention these days because of both short-term effects (negative publicity, low consumer confidence, market share loss, etc.) and long-term effects (stock prices and equity risk). Despite these effects, not many firms are willing to invest in initiatives to decrease disruption risk. According to a study conducted by Computer Sciences Corporation in 2004, 60% of the firms reported that their supply chains are vulnerable to disruption (Poirier and Quinn, 2003). The lack of credible cost/benefit or return on investment (ROI) analyses may be one key reason why companies are not investing in disruption management (Rice and Caniato, 2004; Zsidisin et al., 2001, 2004).

Another survey conducted by CFO Research Services concluded that 38% of 247 companies acknowledged that they had too much unmanaged supply chain risk (cf. Eskew, 2004). While the exact reasons are not known, Rice and Caniato (2003) and Zsidisin et al. (2000) conjecture two key reasons: (1) firms are not familiar with ways to manage supply chain risk; and that (2) firms find it difficult to return on investment analysis to justify certain risk reduction programs or contingency plans.

To garner support for implementing certain risk reduction programs without exact cost/benefit analyses of certain risk reduction programs, effective risk reduction programs must provide strategic value and reduce supply chain risks at the same time (Tang, 2006). Therefore, as articulated by Chopra and Sodhi (2004), the biggest challenge is to determine ways to mitigate supply chain risks and increase profits simultaneously so that companies can achieve a higher level of efficiency by reducing risk while increasing reward – this is also our aim with time-based risk management.

One stream of the risk-mitigation literature focuses on preventing rare risk events. For instance, Lee and Wolfe (2003) describe how different initiatives developed by Homeland Security (e. g., smart containers, Customs-Trade Partnership against Terrorism) would prevent terrorist attacks at various ports in the United States. However, such initiatives may not always be economical for preventing disruptions that are rare and could occur anywhere in a complex supply chain. Our proposed approach aims at preparedness in the supply chain to reduce impact once such an incident occurs.

3.3 Time-Based Risk Management

Akin to various time-based initiatives such as *time-based competition* (cf. Blackburn, 1990; Stalk and Hout, 1991), our "time-based risk management" concept focuses on ***time*** and response processes instead of ***cost***, ***probabilities*** or ***impact***.

Our time-based management concept is based on three elements of time: time to *detect* a disruption ($D1$), time to *design* (or prescribe) a solution in response to the disruption ($D2$), and time to *deploy* the solution ($D3$). After deployment, the time it takes to restore the supply chain operations is the recovery time $R2$. Companies can reduce the impact of supply chain risk incidents by shortening these three elements of time and hence the response time. Increasing responsiveness can help in general and may help increase market competitiveness for the company.

Although there have been efforts to prescribe effective recovery plans for reducing the impact of supply chain disruptions (cf. Chopra and Sodhi, 2004; Lee, 2004; Sheffi, 2005; Tang, 2006), the focus is on recovery after the event has occurred. In contrast, our time based risk management concept focuses on planning and setting up procedures and protocols before a risk event occurs: detection systems and procedures to reduce $D1$, pre-packaged designs to reduce $D2$, and identified communication channels for deployment to reduce $D3$. Just as 80% of the total cost of a product is determined during the product design phase, the activities that take place for designing response can have significant effect on the overall impact of a disruption.

The 3 "D" components of time can be illustrated by using the failed relief efforts associated with Hurricane Katrina. Despite live TV coverage of Katrina's aftermath in late August of 2005, it took three days for FEMA director Michael Brown to learn of those 3000 stranded evacuees at New Orleans' Convention Center. In our terminology, the detect lead time $D1 = 3$ days. According to the Reynolds (2005), communication and coordination between FEMA and local authorities were poor: it took days to sort out who was to do what, when and how. For example, it took two days for Louisiana Governor Blanco to decide on the use of school buses to remove those stranded evacuees. In our context, the design time $D2$ is two days. However, as seen on live TV, most school buses were stranded in the flooded parking lots. FEMA requested over 1000 buses to help out but only a dozen or so arrived the day after; hence, the deploy lead time $D3$ was quite long. As a result, New Orleans has not fully recovered as of this writing; i. e., the recovery time $R2$ exceeds two years. The Katrina fiasco suggests that one can reduce the impact – number of deaths, costs, and recovery time $R2$ – associated with a disruption by reducing the response lead time $R1 = D1 + D2 + D3$.

3.3.1 Response Time and Impact

To understand the importance of shortening the response time, consider the following examples:

Eradicating Med flies in California in 1980. Despite the initial med fly eradication efforts in the mid-70s, med flies were detected in California again in the early part of 1980. Instead of calling for aerial spray of Malathion in a small area (30 square miles) that is proven to be effective but costly, Governor Brown approved the release of sterile male flies and traps. Unfortunately, these methods were not effective and the area of infestation expanded more than 20-fold from 30 square miles in June of 1980 to 620 square miles in July of 1981. As Japan and other countries imposed import restrictions, Governor Brown was under political pressure to approve the aerial spray over an area of 1500 square miles starting July 14, 1981. This delayed action came at a significant cost: an expenditure of over $ 100 million and Governor Brown's political career. See Dawson et al. (1998) and Denardo (2002) for details.

Ground shipping after September 11. Soon after the 9/11 attacks, Chrysler requested their logistics providers to switch the mode of transportation from air to ground. Speedy deployment of this strategy enabled Chrysler to get the parts from such suppliers as TRW via ground transportation. By the time Ford decided to switch to ground shipping, all ground transportation capacity has been taken up and Ford was unable to deploy the same strategy. Facing parts shortages, Ford had to shut down 5 of the US plants for weeks and reduce its production volume by 13% in the fourth quarter of 2001 (cf. Hicks, 2002).

Recovering after supply disruption. Both Ericsson and Nokia were facing supply shortage of a critical cellular phone component (radio frequency chips) after their key supplier, Philip's Electronics semiconductor plant in New Mexico, caught on fire in March of 2000. Nokia recovered quickly by doing the following. First, Nokia immediately sent an executive team to visit Philip's in New Mexico so as to assess the situation. Second, Nokia reconfigured the design of their basic phones so that the modified phones can accept slightly different chips from other Philip's plants; and third, Nokia requested Philip's to produce these alternative chips immediately at other locations. Consequently, Nokia satisfied customer demand smoothly and obtained a stronger market position mainly due to their speedy deployment of their recovery plan. On the contrary, Ericsson was unable to deploy a similar strategy later because all Philip's production capacity at other plants has been taken up by Nokia and other existing customers. Facing with supply delay, Ericsson lost $ 400 million in sales (cf. Hopkins, 2005).

The above examples suggest that the recovery lead time $R2$ is by and large increasing in the response lead time $R1$. This is mainly because the magnitude of the problem triggered by the event escalated exponential over time.

3.3.2 Modeling Disruption Impact over Time

To motivate the impact of a risk incident over time, we consider the epidemiological and the forest fire literature. Specifically, the total impact of a natural disruption (an epidemic or a forest fire) tends to increase super-linearly or even exponentially with time initially and then to taper off. Thus, as shown in Fig. 3.1, shortening the response time $R1$ $(= D1 + D2 + D3)$ can reduce the total recovery time $R2$ and hence the total eventual impact of the disruption.

Epidemic Models. The simplest form of an epidemic model is the *exponential model* that can be explained as follows. Let $I(t)$ be the number of people infected at

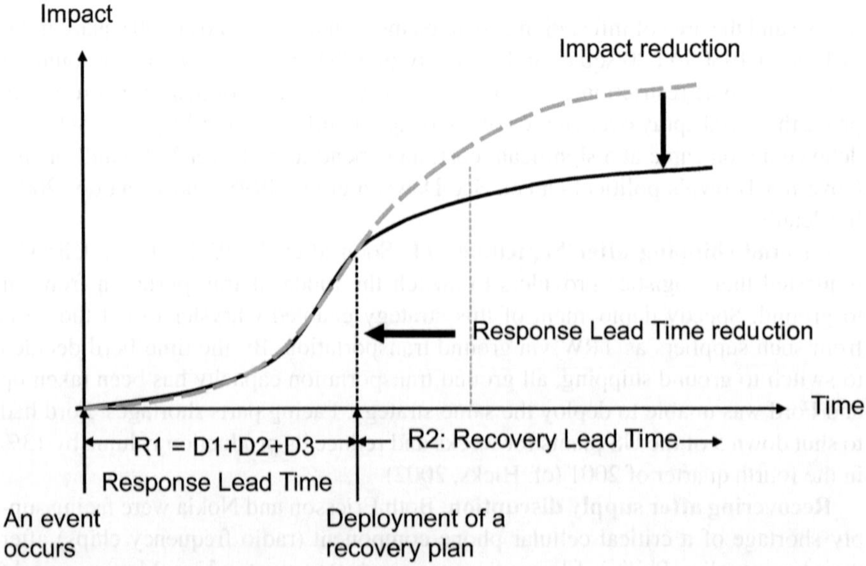

Fig. 3.1 The effect of reducing the response lead time $R1$

time t. In this case, the rate of infection can be defined by the differential equation: $\frac{\mathrm{d}I(t)}{\mathrm{d}t} = k\,I(t)$, where the parameter $k > 0$. This differential equation yields: $I(t) = I_0\,\mathrm{e}^{kt}$, where I_0 is the number of people infected at time 0. Therefore, the number of people infected grows exponentially overtime. By contrast, the *logistic model* is another simple model that stipulates that the infection rate depends on the number of people infected and the number of people who is susceptible to the infection. The number of people infected $I(t)$ grows exponentially initially and plateaus later on (Mollison, 2003).

Fire Impact Models. There are many different types of fire models based on a system of differential equations for estimating the burned areas over time – see, for example, Richards (1995) and Janssens (2000) for various fire spread models. The simplest model is the elliptical fire spread model (cf. Arora and Boer, 2005). that assumes that the burned area takes on the form of an ellipse with the point of ignition at one of the foci. By assuming that the fire is spread linearly over time across this two-dimensional space, they show that the total area burned (size of the ellipse) grows as a squared function of time elapsed since the start of the fire. Thus, the total area burned grows exponentially over time initially, slowing down later as the area of the remaining forest decreases.

In addition to these two models, there are hazards analysis reports highlighting that the magnitude of the problems associated with many hazards (fire, terrorism, earthquake, etc.) tend to grow super-linearly or exponentially over time (Anderson et al., 2002).

While we are not aware of any scientific study of modeling impact over time in the context of supply chain risk, it is plausible that a pattern similar to the logistic model will emerge (Fig. 3.1). For instance, a week's supply delay would have caused little damage to Ericsson but several weeks of supply delay resulted in $ 400 million loss in sales in the first quarter and $ 2 billion eventually in the first year after the dust had settled. Based on such examples, we can point to two ideas for investigation as further research:

(1) The recovery lead time $R2$ and cost can be significant higher if the deployment of a recovery plan is delayed; and
(2) The execution of a recovery plan can become much more difficult if its deployment is delayed.

3.3.3 Time-Based Risk Management in Practice

We now present five time-based disruption management activities that would enable a company to reduce the response time $R1 = D1 + D2 + D3$ (and consequently the recovery time $R2$ and the total impact):

1. **Work with suppliers and customers to map risks**: Many companies already identify potential disruptions according to its impact and likelihood as part of business continuity efforts. They can further that effort in two ways by tracing the impact of each disruption along the supply chain from upstream partners and to downstream customers. Doing so requires discussion among supply-chain partners and therefore creates shared awareness of different types of disruptions and their impacts on different parties. This generates support for collaborative efforts for mitigating risks for all parties.
2. **Define roles and responsibilities**: Companies should work with all key supply chain partners to define the roles and responsibilities to improve communication and coordination when responding to a disruption. Van Wassenhove (2006) suggests three forms of coordination for effective response efforts: (1) coordination by command (central coordination), (2) coordination by consensus (information sharing), and (3) coordination by default (routine communication). The coordination mechanism, agreed among the parties *before* a risk event has taken place, can then be used without further discussion for design and for deployment of solutions *after* the risk event has occurred. Coordination by command seems to be appropriate during the design and deployment phases (i. e., during $D2$ and $D3$). This is because, during these two phases, an identified group within the firm or comprising partners' representatives as well, needs to take central command for collecting and analyzing information to design a recovery plan, and then for disseminating information regarding the deployment of the selected recovery plan. However, to get to this point of agreed upon procedures, coordination by consensus is likely more effective in agreeing on detection mechanisms and in designing pre-packed solutions that anticipate disruptions.

3. **Develop monitoring/advance warning systems for detection**: Companies need to develop mechanisms to discover a disruption quickly when it occurs and/or to predict a disruption before it occurs. They must also identify ways to share the information with their supply-chain partners and to get similar information from them. Monitoring and advance warning systems can enable firms to reduce detection time. For instance, many firms have various IT systems for monitoring the material flows (delivery and sales), information flows (demand forecasts, production schedule, inventory level, quality) along the supply chain on a regular basis. For example, Nike has a "virtual radar screen" for monitoring the supply chain operations (Hartwigsen, 2005); Nokia monitors the delivery schedule of their suppliers (Hopkins, 2005); and Seven-Eleven Japan monitors the production/delivery schedule from their vendors (suppliers) as well as the point of sales data from different stores throughout the day (Lee and Whang, 2006). These monitoring systems typically use control charts and, if any anomaly is detected, issue alerts. Hence, these monitoring systems would reduce the detection time $D1$. Besides monitoring systems, advance warning systems are intended to detect an undesirable event before it actually occurs. For example, Allmendinger and Lombreglia (2005) described how different smart alert systems enabled GE to conduct remote sensing and diagnostics so that it can deploy engineers to service their turbines before failures occur.
4. **Design recovery plans**: Develop contingent recovery plans for different types of disruptions. Establishing contingent recovery plans for different types of disruptions *in advance* would certainly reduce the design time $D2$. Many firms have developed various recovery plans (or contingency plans) in advance. For example, Li and Fung has different contingent supply plans that will enable them to switch from one supplier in one country to another supplier in a different country (St. George, 1998). Also, Seven-Eleven Japan has developed different contingent delivery plans that will allow them to switch from one transportation mode (trucks) to a different transportation mode (motorcycles), depending on the traffic condition (Lee and Whang, 2006).
5. **Develop scenario plans and conduct stress tests**: Companies need to create different scenarios and rehearse different simulation runs/drills based on different scenarios with all key supply chain partners. Because the deployment time $D3$ accounts for the preparation time to launch the selected recovery plan, scenario planning and stress tests are effective mechanisms for reducing $D3$. For example, by rehearsing different response and recovery plans at each P&G site under different scenarios annually, P&G managed to restore the operations of its coffee plant in New Orleans by mid-October 2005 after the Katrina's landfall in late August 2005. P&G attributed their quick recovery ($R1 + R2 = 2$ months) to its readiness (Contrill, 2006).

3.4 Risk and Reward Considerations

Besides reducing disruption risks, time-based disruption management can increase a firm's competitiveness as well. As we mentioned earlier, effective risk reduction programs must provide strategic value and reduce supply chain risks at the same time (Tang, 2006). In other words, we should look for ways to mitigate supply chain risks and increase profits simultaneously so that companies can achieve a higher level of efficiency by reducing risk while increasing reward (Chopra and Sodhi, 2004). Time-based risk management may help achieve this (Fig. 3.2).

We can use Spanish apparel maker Zara's success to illustrate how time-based disruption management can enable firms to increase their competitiveness by responding to dynamic changes quickly (as articulated in the time-based competition literature). Despite the fact that fashion retailers are susceptible to many supply chain risks such as uncertain supply, transportation delays, shrinkage and theft, uncertain demand, Zara continues to be a profitable European fashion retailer with sales and net incoming growing at an annual rate of over 20% as of 2008. Ferdows et al. (2004) attributed Zara's success to its "rapid fire fulfilment" strategy that enables Zara to increase its competitiveness while reducing risks (Fig. 3.2). Specifically, Zara's claim to fame is time reduction: Zara is capable of design, manufacture and ship a new line of clothing within 2 weeks, while most traditional fashion retailer will take more than 24 weeks.

Zara operates as a vertically integrated supply chain by co-locating designers and the factory in Spain, by managing their warehouses and distribution centers in Spain, and by running all Zara stores worldwide (Ghemawat and Nueno, 2003). Not only does this integrated supply chain provide Zara visibility, it also enables Zara

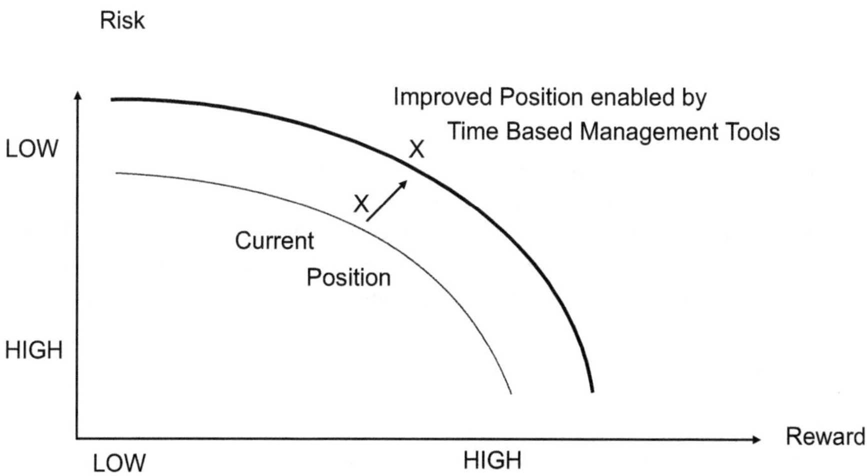

Fig. 3.2 Time Based Disruption Management enables a firm to shift position to a higher efficiency risk-reward curve. Adapted from Chopra and Sodhi (2004)

to facilitate close communication and coordination with all supply chain partners. By receiving point of sales data from their own stores on a regular basis, Zara has a well established process for analyzing the sales data to detect sudden changes in demand and/or fashion trends.

As such, Zara's detect lead time $D1$ is short. By managing centrally and by working closely with all partners, Zara can analyze the situation and prescribe a response should the market change suddenly. By co-locating the designers and the factory, Zara has the capability to deploy different recovery plans by designing and manufacturing new designed clothes very quickly should the sales of existing designs are below expectations. Hence, the design time $D2$ is short. The close proximity of suppliers, designers, factories, and distribution centres enable Zara to communicate, coordinate, and deploy the selected recovery plan quickly with all supply chain partners (from suppliers to the stores). Hence, the deploy lead time $D3$ is also short. In addition, Zara also has capability to implement recovery plans quickly so as to reduce the deploy time $D3$ and recovery time $R2$: for example, Zara engages in "flexible supply contracts" with "multiple suppliers" to be able to adjust the order quantity quickly should a demand disruption occur.

Not only do these mechanisms help Zara to reduce $D1$, $D2$, $D3$ and consequently $R2$, they enable Zara to increase its competitiveness and profit by: (a) generating more accurate demand forecasts in a timely manner; (b) designing, producing, and distributing newly designed products in small batches quickly; and (c) reducing the costs associated with price markdowns (due to over-stocking) and lost sales (due to under-stocking). Relative to its competitors, the time-based management concept has enabled Zara to achieve higher profitable growth and lower supply chain disruption risk simultaneously.

3.5 Conclusion

We have used diverse examples and natural disruption models (epidemic model and fire model) and anecdotal evidence to argue that firms can use the time-based management concept to reducing the response lead time and recovery lead time, which will in turn reduce the impact of a disruption. We suggested five activities that would enable a firm to reduce the response lead time $R1$. Finally, we suggested using Zara's example that we may be able to achieve both risk reduction and extra rewards through increased responsiveness enabled by time-based risk management.

However, we have presented a concept that requires further research. For instance, we presented impact models from the epidemiology literature – we need to develop similar models for supply chains. Likewise, we need empirical research to study how and to what extent companies are extending business continuity efforts to respond to risk events. We believe time-based risk management to be a rich area for modeling and empirical research.

References

Allmendinger, G., and R. Lombreglia, "Four Strategies for the Age of Smart Services", Harvard Business Review, pp. 1–10, October, 2005

Anderson, C., Hazards and Vulnerability Analysis, Report, University of North Texas, 2002. (Available at: http://training.fema.gov/EMIWeb/downloads/UNTServiceLearning.pdf)

Arora, V.K. and G.J. Boer, "Fire as an Interactive Component of Dynamic Vegetation Models", *Journal of Geophysics Research*, Vol. 110, pp. 1–20, 2005

Billington, C., "HP Cuts Risk with Portfolio Approach", *Purchasing Magazine Online*, February 21, 2002

Blackburn, J., *Time-Based Competition: The Next Battleground in American Manufacturing*, McGraw-Hill Publishers, Massachusetts, 1990

Casey, N. and A. Pasztor, "Safety Agency, Mattel Clash Over Disclosures", *Wall Street Journal*, A.1, September 4, 2007

Chopra S. and M Sodhi, "Avoiding Supply Chain Breakdown", *Sloan Management Review*, Fall, pp. 53–62, Vol. 46, No. 1, 2004

Cottrill, K., "A Game Plan for Disaster Recovery", *MIT Supply Chain Strategy*, Vol. 2, No. 6, pp. 4–9, June, 2006

Dawson, A., S. Hassenpflug, and J. Sloan, "California Agriculture Trade: Combating the Med fly Menace", Case Study, Center for Trade and Commercial Diplomacy, Monterey Institute of International Studies, Monterey, California, 1998

Denardo, E.V., *The Science of Decision Making: A Problem-Based Approach Using Excel*, Wiley Publisher, New York, 2002

Eskew, M., "Mitigating Supply Chain Risk", *CEO*, pp. 25–26, April, 2004

Ferdows, K., M. Lewis, and J. Machuca, "Rapid Fire Fulfilment", *Harvard Business Review*, pp. 104–117, November, 2004

Ghemawat, P. and J. Nueno, "Zara: Fast Fashion", *Harvard Business School Case* # 9-703-497, Harvard Business School, 2003

Hartwigsen, K., "Integrated Crisis Management: Building a framework for unified response", Nike Internal Presentation, 2006

Hendricks, K. and V. Singhal, "An Empirical Analysis of the Effect of Supply Chain Disruptions on Long-run Stock Price Performance and Equity Risk of the Firm", *Production and Operations Management*, pp. 25–53, Spring, 2005

Hicks, M., "When Supply Chain Snaps", *eWeek*, February 18, 2002

Hopkins, K., "Value Opportunity Three: Improving the Ability to Fulfill Demand", *Business Week*, January 13, 2005

Janssens, M., *Introduction to Mathematical Fire Modeling*, Second Edition, CRC Publisher, 2000

Lee, H., "The Triple-A Supply Chain", *Harvard Business Review*, pp. 102–112, October, 2004

Lee, H. and M. Wolfe, "Supply Chain Security Without Tears", Supply Chain Management Review, Vol. 7, No. 1, pp. 12–20, January, 2003

Lee, H, and J. Whang, "Seven Eleven Japan", *Stanford Business School case* # GS-18, 2006

Martha J., and S. Subbakrishna, "Targeting a Just-in-Case Supply Chain for the Inevitable Next Disaster", *Supply Chain Management Review*, September, 2002

Mollison, D., *Epidemic Models: Their Structure and Relation to Data*, Cambridge University Press, 2003

Poirier, C. and F. Quinn, *Calibration Supply Chain Management*, Computer Sciences Corporation Report, 2003

Pyke, D. and C.S. Tang, "How to Mitigate Product Safety Risks Proactively? Process, Challenges and Opportunities", working paper, UCLA Anderson School, 2008

Reynolds, P., "Multiple Failures Caused Relief Crisis", BBC News, September 7, 2005. (Article available at: http://news.bbc.co.uk/1/hi/world/americas/4216508.stm)

Rice B. and Caniato F., "Supply Chain Response to Terrorism: Creating Resilient and Secure Supply Chains", Supply Chain Response to Terrorism Project Interim Report, MIT Center for Transportation and Logistics, MIT, Massachusetts, 2004

Richards, G.D., "A general mathematical framework for modelling two dimensional wildland fire spread". *International Journal of Wild land Fire*, **5**, 63–72, 1995

Sheffi, Y., "Building a Resilient Supply Chain", *Harvard Business Review*, pp. 3–5, October, 2005

St. George, A., "Li, and Fung: Beyond 'Filling in the Mosaic' ", *Harvard Business School* case 9-398-092, 1998

Stalk, G. and T. Hout, *Competing Against Time: How Time-based Competition is Reshaping Global Markets*, Free Press, New York, 1991

Tang, C.S. "Robust Strategies for Mitigating Supply Chain Disruptions", *International Journal of Logistics: Research and Applications*, Vol. 9, pp. 33–45, 2006

Van Wassenhove, L., "Humanitarian Aid Logistics: Supply Chain Management in High Gear", *Journal of Operational Research Society*, Vol. 57, pp. 475–489, 2006

Zsidisin, G., Panelli, A., and Upton, R. "Purchasing Organization Involvement in Risk Assessments, Contingency Plans, and Risk Management: An Exploratory Study", *Supply Chain Management: An International Journal*, pp. 187–197, 2001

Zsidisin, G., Ellram, L., Carter, J., and Cavinato, J. "An Analysis of Supply Risk Assessment Techniques", International Journal of Physical Distribution and Logistics Management, Vol. 34, No. 5, pp. 397–413, 2004

Chapter 4
Prioritization of Risks in Supply Chains

Mohd. Nishat Faisal

Abstract Modern supply chains are very complex, with physical, financial, and information flows occurring simultaneously in order to ensure that products are delivered in the right quantities, to the right place in a cost-effective manner. Maintaining uninterrupted supply chain flows is a prerequisite for the success of a supply chain in the marketplace. But there are always associated risks in each of these flows which require suitable strategies to mitigate them. The issue of risks in supply chains has assumed importance in wake of the understanding that supply chain failures are fatal to the existence of all the partners' in a supply chain. The severity of supply chain failures are more felt by small and medium enterprises (SMEs) who form the majority at tier II and tier III levels of a supply chain. This is because of the limited resources and lack of adequate planning to counter supply chain risks. Management of risk in supply chains is a multi-criteria decision making problem. The research presented in this chapter proposes a Fuzzy-AHP based framework to prioritize various risks in supply chains. An exhaustive literature review complimented with the experts' opinion was undertaken from the perspective of SMEs to formulate a hierarchical structure of risks in supply chains. A fuzzy analytic hierarchical process (F-AHP) is then utilized to ascertain the relative weightings which are subsequently used to prioritize these risks. Understanding the priorities would help the firms to accord importance and develop suitable strategies to manage supply chain risks according to their relative importance. This provides effective management of scarce resources available to SMEs to manage risks resident in their supply chains.

4.1 Introduction

In today's interconnected and information based economies, there have been dramatic shifts in the way companies interact, driven by both new technologies and new business models. This has led to firms' exposure to new forms of risks, many

T. Wu, J. Blackhurst (eds.), *Managing Supply Chain Risk and Vulnerability*

related to their extended supply chains. Though firms' recognize that the survival in the modern business environment is no longer an issue of one firm competing against another firm but has, instead, become an issue of one supply chain competing against another supply chain the understanding that failure of any one link in the supply chain would have ripple effect on the overall supply chain has come rather late. But there is a growing realization among the companies that adopting supply chain risk management practices can yield continuous improvement of supply chain operations (Elkins et al., 2005).

The risk in a supply chain is the potential variation of outcomes that influence the decrease of value added at any activity cell in a chain, where the outcome is described by the volume and quality of goods in any location and time in a supply chain flow (Bogataj and Bogataj, 2007). The consequences of failing to manage risk have also become more severe. In addition to the direct impact on revenue and profit, disruptions in supply or demand can hurt a firm's trading partners (e.g. customers and suppliers), since the interconnectedness of a supply chain has a ripple effect that affects the entire supply chain ecosystem (Shi, 2004). It is also reported that companies experiencing such disruptions under-perform their peers significantly in stock performance as well as in operating performance as reflected in costs, sales, and profits (Hendricks and Singhal, 2003, 2005). Thus, management of risk is, or should be, a core issue in the planning and management of any organization (Finch, 2004).

All the parts of supply chain could be impacted by a great variety of risks and the supply chain risks can have significant impact on the firm's short-term and long-term performance. For the success of supply chain, the method to find ways of mitigating supply chain risks is critical to manage supply chain in unstable environment. In industries moving towards seamless supply chains the issue of supply chain risk handling and risk sharing along the supply chain is a topic of growing importance (Agrell, 2004; Gunasekaran et al., 2004). A key feature of supply chain risk is that, by definition, it extends beyond the boundaries of the single firm and, moreover, the boundary spanning flows can become a source of supply chain risks (Jüttner, 2005). Thus, to assess supply chain risk exposures, companies must identify not only direct risks to their operations, but also the potential causes or sources of those risks at every significant link along the supply chain (Christopher et al., 2002; Souter, 2000). The reasons that make an integrated approach to supply chain risks analysis and management important are (Harland, 2003; Shi, 2004; Faisal et al., 2006):

- Examining risk factors in isolation makes it difficult to understand their interactions.
- There may be an increase in risk management costs, since firms may unnecessarily hedge certain risks that are in reality offset by others.
- A fragmented approach to risk management also increases the likelihood of ignoring important risks.
- ICT revolution has eliminated the geographical boundaries for developing supply chain partnerships.

- Even for known risks, it is important to consider their overall impact to the entire organization. Otherwise mitigation attempts may only introduce new risks, or shift the risk to less visible parts of the organization.
- Failure to consider risk interactions can also cause firms to grossly underestimate their risk exposures.

There has been a recent surge in interest and publications from academics and practitioners regarding supply chain disruptions and related issues as supply chain risks can potentially be harmful and costly for the whole supply chain (Craighead et al., 2007). One important constituent of risk management process is the prioritization of risks. Prioritization helps a company to focus the decision making and risk management effort on the most important risks (Hallikas et al., 2002). Prioritization requires comparisons concerning the relative importance of each of the risk variables.

The purpose of this chapter is to contribute and provide a more complete understanding of risk in supply chains. It seeks to develop a model to assess relative importance of numerous risks inherent in the supply chain. The primary aim is to illustrate how organizations can prioritize risks in their supply chains. The main research problems addressed by the study presented in this chapter are:

(1) What kinds of risks are associated with various supply chain flows?
(2) How these risks can be prioritized?

After introduction, the remainder of this chapter is organized as follows. Sect. 4.2 presents literature review on risk, supply chain risk management, and risk mitigation strategies for supply chains. Risk taxonomy for supply chain is presented in Sect. 4.3. Next, Sect. 4.4 of the chapter discusses the methodology for prioritizing supply chain risks. A graphical representation of the model is shown in Fig. 4.1. In this part the numerical application of the proposed model for small and medium enterprises (SMEs) cluster is also discussed. Finally, Sect. 4.5 presents the concluding remarks of the chapter.

4.2 Literature Review

4.2.1 Risk

On a very general level, risk is defined as the probability of variance in an expected outcome and it differs from uncertainty in that risk has associated with it a probability of a loss and uncertainty is an exogenous disturbance (Spekman and Davis, 2004). According to Norrman and Jansson (2004) "risk is the chance, in quantitative terms, of a defined hazard occurring. It therefore combines a probabilistic measure of the occurrence of the primary event(s) with a measure of the consequences of that/those event(s)". So to manage risk both an assessment of the probability of risk and its impact is necessary (Hallikas et al., 2002; Zsidisin et al., 2004; Hallikas et al., 2004).

As companies increasingly move towards inter-firm co-operation to achieve sustained competitive advantage, research in risk management began to examine risk management at the level of inter-organizational relationships and more recently at the level of supply chains and networks (Harland et al., 2003). Risk is perceived to exist when there is a relatively high likelihood that a detrimental event can occur and that event has a significant associated impact or cost (Zsidisin et al., 2004). A key feature of supply chain risk is that, the complexity makes it difficult for the exposed company to estimate the total financial losses, which contributes to the impediment in how to design risk mitigation solutions for supply chains. Current business trends that increase the vulnerability to risks in supply chains are (Harland et al., 2003; Normann and Jansson, 2004; Christopher and Lee, 2004; Cucchiella and Gastaldi, 2006):

- increased use of outsourcing of manufacturing and R&D to suppliers;
- globalization of supply chains;
- reduction of supplier base;
- more intertwined and integrated processes between companies;
- reduced buffers;
- shorter lead times requirements;
- shorter product life cycles and compressed time-to-market;
- increased product/service complexity; and
- capacity limitation of key components.

4.2.2 Supply Chain Risk Management

The degree of the vulnerability of a supply chain is determined to a large extent by the degree of complexity of the network (Nieger et al., 2008). In recent times the complexity has increased many-fold due to firms' focus on their core competence and increased dependence on outsourcing. Top executives at Global 1000 firms now consider supply chain disruptions and their associated operational and financial risks to be their single most pressing concern (Craighead et al., 2007). Risk management in supply chain cannot be equated to disaster response. Rather, it means keeping an increasingly complex process moving efficiently at the lowest total cost and without compromising the quality of the product or customer satisfaction (Hauser, 2003). Supply chain risk management (SCRM) is defined as "the process of risk mitigation achieved through collaboration, co-ordination and application of risk management tools among the partners, to ensure continuity coupled with long term profitability of the supply chain" (Faisal et al., 2007a). SCRM is still a fairly new field of research and studies related to the topic are scarce (Ojala and Hallikas, 2006; Jüttner, 2005). It should also be noted that risks cannot be completely eliminated from supply chains but strategies can be developed to manage these risks if the dynamics between the variables related to risks in a supply chain are understood (Faisal et al., 2006).

It is important for supply chain managers to recognize that in taking action to reduce known risks, they are changing the risk profile for that organization and for others in the network (Peck, 2005). Thus, for mitigating risk in supply chains it is required to expend the risk management focus from the companies' own sites to suppliers and sub-suppliers. There is a need to work together in risk identification, assessment, management and business continuity planning and also of formal assessment of how suppliers are working with those issues and by putting requirements into contracts (Normann and Jansson, 2004). Risk management skills which includes, awareness of risk signals, developing risk management plans, and improving end to end information visibility are essential requirements for supply chain management success (Giunipero and Pearcy, 2000; Christopher and Lee, 2004). While revisiting single-sourcing decisions and changing inventory management policies will likely help maintain continuity during future crises, experts have clearly demonstrated, and logistics managers have candidly admitted, that firms need to vastly improve their disaster management planning for managing risk in supply chains (Hale and Moberg, 2005).

4.2.3 Supply Chain Risk Mitigation Strategies

A risk management system is basically an action plan that specifies which risks can be addressed, and how to address them (Shi, 2004). Elkins et al. (2005) suggest 18 best practices to mitigate supply chain risks. Companies may not need to implement all 18 of the best practices to improve supply chain risk management capabilities. Rather, they should prioritize these 18 best practices as some of these actions can be taken with a minimal level of investment and should yield immediate benefits. There should be a plan that identifies the short-term actions that can he deployed with a minimum of investment and establish a roadmap for deploying intensive project-team resources, business intelligence systems, and improved supply chain infrastructure. SAM framework (Kleindorfer and Saad, 2005) for supply chain risk management comprise three main tasks that have to be practiced continuously and concurrently as the foundation of disruption risk management. The three tasks are: Specifying sources of risk and vulnerabilities, Assessment, and Mitigation (SAM). Further, to implement the three SAM tasks introduced above, the authors have formulated a set of 10 principles, derived from the industrial risk management and supply chain literatures. Table 4.1 summarizes nine different robust supply chain strategies that aim to improve a firm's capability to better manage supply and/or demand under normal circumstances and to enhance a firm's capability to sustain its operations under risk (Tang, 2006).

With so many related risks and risk-mitigation approaches to consider, managers must do two things when they begin to construct a supply-chain risk management strategy. First, they must create a shared, organization-wide understanding of supply-chain risk. Then they must determine how to adapt general risk-mitigation approaches to the circumstances of their particular company. Managers can achieve

Table 4.1 Risk mitigation strategies in supply chains

Robust Supply Chain Strategy	Main Objective	Benefit(s) under normal circumstances	Benefit(s) after a major disruption
Postponement	Increases product flexibility	Improves capability to manage supply	Enables a firm to change the configuration of different products quickly
Strategic Stock	Increases product availability	Improves capability to manage supply	Enables a firm to respond to market demand quickly during a major disruption
Flexible Supply Base	Increases supply flexibility	Improves capability to manage supply	Enables a firm to shift production among suppliers promptly
Make-and-Buy	Increases supply flexibility	Improves capability to manage supply	Enables a firm to shift production between in-hour production facility and suppliers rapidly
Economic Supply Incentives	Increases product availability	Improves capability to manage supply	Enables a firm to adjust order quantities quickly
Flexible Transportation	Increases flexibility in transportation	Improves capability to manage supply	Enables a firm to change the mode of transportation rapidly
Revenue Management	Increases control of product demand	Improves capability to manage demand	Enables a firm to influence the customer product selection dynamically
Dynamic Assortment Planning	Increases control of product demand	Improves capability to manage demand	Enables a firm to influence the demands of different products quickly
Silent Product Rollover	Increases control of product exposure to customers	Improves capability to manage supply and demand	Enables a firm to manage the demands of different products swiftly

Adapted from Tang (2006)

the former through stress testing and the latter through tailoring risk management approaches (Chopra and Sodhi, 2004).

4.3 Supply Chain Risks Taxonomy

Supply chain structure is made up of interdependent parts that come together to provide the required products and services to the satisfaction of the customers.

Tied closely to this interdependence is the role of uncertainty within the supply chain (Sounderpandian et al., 2008). The literature in the risk management field indicates that the primary sources of risk to the business organization may be categorized into exogenous and endogenous (Ritchie and Brindley, 2000). In a supply chain, risks can be classified into two types: risks arising from within the supply chain network and risks external to it. For the former, the attributes are due to the interaction between firms across the entire supply chain network. This set of internal risks can encompass supply risk, demand risk, and trade credit risk for instance. External risks, on the other hand, arise from the interactions between the supply chain network and its environment, such as international terrorism, and natural disasters like SARS (Goh et al., 2007). Jüttner et al. (2002) suggest organizing risk sources relevant for supply chains into three categories: external to the supply chain, internal to the supply chain, and network related.

Following Cavinato (2004), Spekman and Davis (2004), and Jüttner (2005) risks in supply chains are classified under four sub-chains, physical, financial, informational, and relational. Physical sub-chain, represent traditionally viewed logistics, in the form of transportation, warehousing, handling, processing, manufacturing, and other forms of utility activities. Financial sub-chain working in parallel deals with the supply chain's flow of money while informational sub-chains parallel the physical and financial chains through the processes and electronic systems used for creating events and triggered product movements and service mobilization. Relational sub-chains relate to the chosen linkages between buyers, sellers, and logistics parties in between them.

4.3.1 Risks in Physical Sub-Chain

Risks in the actual movements and flows within and between firms, transportation, service mobilization, delivery movement, storage, and inventories can be termed as the risk in physical flow of the supply chain (Cavinato, 2004). Traditionally, most of the earlier definitions of risk in supply chains found in literature took only the physical flow of goods into consideration (Spekman and Davis, 2004). Many of the risk in the physical flow are difficult to anticipate and consequently formal risk management approaches fails to mitigate them. Prominent among this class are those risk which have a low probability of occurrence but high impact like disruption risks. Supplier capacity constraint is a form of risk in physical sub-chain. Constraints exist that restrict a supplier's ability to make rapid changes to varying demands (Zsidisin et al., 2000). Risk in physical sub-chain can also exist in form of fluctuations of demand, such as those caused by the "bullwhip effect" and may tax a supplier beyond its abilities. A supplier may not have extra equipment, available employees, or the ability to obtain necessary inputs to handle rapid spurts in demand. On the other hand, it may also be difficult for suppliers to utilize excess "slack" during order declines, which makes it difficult to attain profits from excess capacity. Physical sub-chain is also impacted by quality-related risks that can cause significant detrimental effects on the purchasing organization, with a cascading effect through the supply chain to the final consumers. Each link within a supply chain is dependent on the other links to meet product or service requirements. Quality failures can stem from the failure of suppliers to maintain capital equipment, lack of supplier training in quality principles and techniques, and damage that occurs in transit (Zsidisin et al., 2000). Risks in physical sub-chain may be further classified as shown Table 4.2.

Table 4.2 Categories of risk in physical sub-chain

SN	Type of risk	Reference(s)	Remarks
1.	Delays (DL)	Chopra and Sodhi (2004)	Due to high utilization or another cause of inflexibility of suppliers
2.	Disruptions (DS)	Chopra and Sodhi (2004); Finch (2004); Johnson (2001); Hale and Moberg (2005); Peck (2005)	Very unpredictable but of high impact.
3.	Supplier capacity constraints (CC)	Zsidisin et al. (2000); Giunipero and Eltantawy (2004)	Unable to handle sudden spurt or to utilize excess slack
4.	Production technological changes (TC)	Zsidisin et al. (2000); Giunipero and Eltantawy (2004)	Supplier not able to produce items to necessary demand level and at a competitive price
5.	Transportation (TR)	Cavinato (2004); Speckman and Davis (2004); Svensson, (2004); Peck (2005)	Pertinent in case of logistics outsourcing as is the case of 3PL
6.	Inventory (IN)	Chopra and Sodhi (2004)	Excess inventory for products with high value or short life cycles can get expensive
7.	Procurement (PR)	Hallikas et al. (2002); Chopra and Sodhi (2004)	Unanticipated increases in acquisition costs
8.	Capacity Inflexibility (CI)	Chopra and Sodhi (2004); Johnson (2001); Giunipero and Eltantawy (2004); Svensson (2004)	Facility fails to respond to changes in demand
9.	Design (DG)	Zsidisin et al. (2000); Speckman and Davis (2004)	Suppliers' inability to incorporate design changes
10.	Poor quality (PQ)	Treleven and Schweikhart (1988); Zsidisin et al. (2000); Svensson (2004); Sounderpandian et al. (2008)	If suppliers plants don't have quality focus

4.3.2 Risks in Financial Sub-Chain

Risks in supply chains due to the flows of cash between organizations, incurrence of expenses, and use of investments for the entire chain/network, settlements, A/R (accounts receivables) and A/P (accounts payables) processes and systems can be classified under financial risks. Financial risks also include factors like settlement process disruption, improper investments, and not bringing cost transparency to the overall supply chain (Cavinato, 2004). In case of strategic alliances the risk of financial stability of alliance partners is crucial for supply chain success (Giunipero and Eltantawy, 2004). According to Peck (2005), managing risks in financial sub-chain is not an easy task as disruptions due to macroeconomic vacillations like currency fluctuations or due to strained relations among nations are difficult to predict and are beyond the direct control of supply chain managers and business strategists.

A better understanding of the causal relationship between supply chain performance and financial measures is critical to both supply chain and financial managers (Tsai, 2008). Financial risks can also present themselves through the risk of reworking stock and penalties for non-delivery of goods (Christopher and Lee, 2004). When a firm has operations in multiple countries, changes in foreign exchange rates can have a significant impact on both the income statement and the balance sheet. Price changes for commodities such as oil and electricity can have an impact on supply chain costs and price changes for commodities like steel and copper can affect the cost of goods sold. Labor-intensive enterprises in many developing countries have felt great pressure from the rise of labor-related costs, such as wages and costs to improve working conditions. A January 2006 report by American Chamber of Commerce in China found that rising labor costs significantly decreased margins in 48% of U.S. manufacturers in China (Jiang et al., 2007). Changing economic scenario in key markets are also a major source of risk in financial sub-chain e.g. recent recession in US economy has severely dented the profit margins of many of the firms in India that used to have major chunk of their products being exported to US markets. Major risks in financial sub-chain are represented in Table 4.3.

4.3.3 Risks in Relational Sub-Chain

This dimension of risk concerns the degree of interdependence among partners and the tendency of a partner to act in its own self interest to the detriment of other supply chain members (Spekman and Davis, 2004). Supply chains are entities of interdependent parts which come together to provide the product /service to the final customer. The type of the relationship is likely to have an effect on the supply chain risks (Ojala and Hallikas, 2006). Though in supply chain literature long term collaborative relationships are recommended (Mentzer et al., 2000; Barratt and Oliveira, 2001; Callioni and Billington, 2001), companies particularly SMEs competing on the basis of low cost depend on short term cost based relationships with their part-

Table 4.3 Categories of risk in physical sub-chain

SN	Type of risk	Reference(s)	Remarks
1.	Cost/price risk (CR)	Treleven and Schweikhart (1988); Zsidisin et al. (2000)	Concerns competitive cost risk
2.	Business risks (BR)	Zsidisin et al. (2000)	Concerns financial stability of the supplier
3.	Fiscal risks (FR)	Harland et al. (2003)	Arises through changes in taxation
4.	Untimely payments (UT)	Speckman and Davis (2004); Shi (2004)	Loss of goodwill and may impact much on SMEs
5.	Settlement process disruption (SP)	Cavinato (2004)	Leads to delay in payments and impacts SC profitability
6.	Volatile oil prices (OP)	Faisal et al. (2006)	Impacts inbound and outbound costs
7.	Lack of hedging (LH)	Speckman and Davis (2004)	Disastrous in case of bankruptcy of partners in the SC
8.	Investment risks (IR)	Hallikas et al. (2002); Ojala and Hallikas (2006)	Caused by economic fluctuations in the market
9.	Unstable pricing (UP)	Speckman and Davis (2004)	May lead to lack of trust among SC partners
10.	Exchange rate risks/ currency fluctuations (ER)	Chopra and Sodhi (2004); Peck (2005)	Impacts the procurement strategy of the firm

ners. In cost based relationships the firms do not share information and partners are switched frequently thus giving rise to risks in relational sub-chains as enunciated below.

4.3.3.1 Reputational Risk (RR)

Outsourcing from low cost destinations is fraught with dangers of associated risks arising from moral issues like child labor, labor health, safety, and welfare in developing countries. Consumers are shunning products that contain materials manufactured under sweatshop labor conditions (Jiang et al., 2007). The cases of labor abuse in the sporting goods and apparel industry, recent being that of Primark show that such negative publicity can damage brands and erode market positions substantially (Zadek, 2004; Harland et al., 2003; Frenkel and Scott, 2002).

4.3.3.2 Lack of Trust and Opportunism Risk (TOR)

One of the most important factors affecting the entire process of supply chain management is the trust among the trading partners (Sinha et al., 2004). Trust may concern a partner's willingness to perform according to agreements, or the intention to do so. Risks exist if the party is not competent to act or if the party chooses not to act (Spekman and Davis, 2004). Lack of trust may also lead to opportunism, where one supply chain partner acts in its own self-interest to the detriment of others.

4.3.3.3 Legal Risk (LR)

These risks expose the firms to litigation with action arising from customers, suppliers, shareholders, and employees (Harland et al., 2003).

4.3.3.4 Intellectual Property Rights Risk (IPR)

Increased reliance on outsourcing creates a loss of control and a risk of losing proprietary information shared between parties (Giunipero and Eltantawy, 2004). If there are no predefined rules and the organizations are not careful regarding the dissemination of information to its suppliers, the probability of losing the proprietary information is high. Today suppliers may work for different organizations at the same time and in general the enforcement mechanisms related to intellectual property are weak in many developing countries and thus there is a risk that proprietary information may be leaked to competitors.

4.3.4 Risks in Informational Sub-Chain

Risks associated with materials flows are not unrelated to the risks associated with information flows. Orders that are transmitted incorrectly, a lack of transparency and visibility in the supply chain, or hesitancy in sharing accurate and timely information with partners, all contribute to a supply chain's inability to perform as intended (Spekman and Davis, 2004). So based on the type of impact that different information risks have on the supply chain, they can be broadly classified as (Faisal et al., 2007b):

4.3.4.1 Information Security/Breakdown Risks (SR)

Today computer based information systems are central to the supply chain and thus their failure can result in a substantial cost. In general this cost can be immediate lost sales, emergency service cost, cost of restoring data, and long-term loss of customer goodwill (Cardinali, 1998). Security is defined as the protection of data against accidental or intentional disclosure to unauthorized persons, or unauthorized modifications or destruction. It is a careful balance between information safeguard and user access (McFadden, 1997). Information security risks arise from hackers, viruses and worms, distributed denial of service attacks or even the internal employee frauds of the organization. Terrorist attacks like 9/11 and natural disasters like Hurricanes, Tsunami have made organizations to rethink their information security strategies.

4.3.4.2 Forecast Risks (FR)

Forecast risks results from a mismatch between a company's projections and actual demand (Chopra and Sodhi, 2005). All kinds of information distortions in a supply chain, often lead to the risks of bullwhip (Piplani and Fu, 2005). It creates situations where the orders to the supplier tend to have larger fluctuations than sales to the customer.

4.3.4.3 Information Systems/Information Technology Outsourcing Risk (OR)

IS/IT outsourcing is broadly defined as a decision taken by an organization to contract-out or sell the organization's IT assets, people and/or activities to a third party supplier, who in exchange provides and manages assets and services for monetary returns over an agreed time period (Kern and Willcocks, 2000). IT outsourcing risks include opportunism by vendors, hidden costs, loss of control, poaching, and information security apprehensions.

4.4 Methodology

The analytic hierarchy process (AHP) developed by Saaty (1980) is a multi-criteria decision-making tool that can handle unstructured or semi-structured decisions with multi-person and multi-criteria inputs. It also allows users to structure complex problems in the form of a hierarchy or a set of integrated levels. In addition to this, AHP is easier to understand and can effectively handle both qualitative and quantitative data (Durán and Aguilo, 2008). The AHP attracted the interest of many researchers for long because of its easy applicability and interesting mathematical properties (Sasamal and Ramanjaneyulu, 2008). AHP involves the principles of decomposition, pair wise comparisons, and priority vector generation and synthesis.

Though the purpose of AHP is to capture the expert's knowledge, the conventional AHP still cannot reflect the human thinking style. In spite of its popularity, this method is often criticized because of a series of pitfalls associated with the AHP technique which can be summarized as follows (Durán and Aguilo, 2008):

- Its inability to adequately handle the inherent uncertainty and imprecision associated with the mapping of the decision-maker's perception to exact numbers.
- In the traditional formulation of the AHP, human's judgments are represented as exact (or crisp, according to the fuzzy logic terminology) numbers. However, in many practical cases the human preference model is uncertain and decision-

makers might be reluctant or unable to assign exact numerical values to the comparison judgments.
- Although the use of the discrete scale of 1–9 has the advantage of simplicity, the AHP does not take into account the uncertainty associated with the mapping of one's judgment to a number.

A good decision-making models needs to tolerate vagueness or ambiguity since fuzziness and vagueness are common characteristics in many decision-making problems (Yu, 2002). Due to the fact that uncertainty should be considered in some or all of the pairwise comparison values, the pairwise comparison under traditional AHP, which needs to select arbitrary values in the process, may not be appropriate (Yu, 2002). Thus the use of fuzzy numbers and linguistic terms may be more suitable, and the fuzzy theory in AHP should be more appropriate and effective than traditional AHP in an uncertain pairwise comparison environment (Kang and Lee, 2007). Fuzzy set theory bears a resemblance to the logical behavior of human brain when faced with imprecision (Cakir and Canbolat, 2008). It has the advantage of mathematically representing uncertainty and vagueness and provides formalized tools for dealing with the imprecision intrinsic to many problems (Chan and Kumar, 2007).

Although the purpose of the original AHP was to capture expert knowledge, conventional AHP did not truly reflect human cognitive processes-especially in the context of problems that were not fully defined and/or problems involving uncertain data (so-called "fuzzy" problems) (Fu et al., 2006). Laarhoven and Pedrycz (1983) therefore introduced the concept of "fuzzy theory" to AHP assessments. This so-called "fuzzy analytic hierarchical process" (fuzzy AHP) was able to solve uncertain "fuzzy" problems and to rank excluded factors according to their weight ratios. The research presented in this chapter prefers Chang's extent analysis method (Chang 1992, 1996) since the steps of this approach are relatively easier than the other fuzzy AHP approaches and similar to the conventional AHP (Büyüközkan et al., 2008; Bozbura et al., 2007; Büyüközkan, 2008; Chan and Kumar, 2007). Fuzzy-AHP has been widely used for prioritization purposes like prioritization of organizational capital (Bozbura and Beskese, 2007), key capabilities in technology management (Erensal et al., 2006), prioritization of human capital measurement indicators (Bozbura et al., 2007), adoption of electronic marketplaces (Fu et al., 2006), and supply base reduction (Sarkar and Mohapatra, 2006).

A fuzzy number is a special fuzzy set $F = \{x \in R|\mu_F(x)\}$, where x takes its values on the real line $R_1 : -\infty < x < +\infty$ and $\mu_F(x)$ is a continuous mapping from R_1 to the close interval [0, 1]. A triangular fuzzy number can be denoted as $M = (l, m, u)$. Its membership function $\mu_M(x) : R \to [0, 1]$ is equal to:

$$\mu_M(x) = \begin{cases} 0\,, & x < l \text{ or } x > u\,, \\ (x-l)/(m-l)\,, & l \leq x \leq m\,, \\ (x-u)/(m-u)\,, & m \leq x \leq u\,. \end{cases} \tag{4.1}$$

Where $l \leq m \leq u$, l and u stand for the lower and upper value of the support of M, respectively, and m is the mind-value of M. When $l = m = u$, it is a non fuzzy number by convention. The main operational laws for two triangular fuzzy numbers

M_1 and M_2 are as follows (Kauffman and Gupta, 1991):

$$\begin{aligned} M_1 + M_2 &= (l_1 + l_2, m_1 + m_2, u_1 + u_2)\ , \\ M_1 \otimes M_2 &\approx (l_1 l_2, m_1 m_2, u_1 u_2)\ , \\ \lambda \otimes M_1 &= (\lambda l_1, \lambda m_1, \lambda u_1), \lambda > 0, \lambda \\ M_1^{-1} &\approx (1/u_1, 1/m_1, 1/l_1)\ . \end{aligned} \tag{4.2}$$

Let $X = \{x_1, x_2, \ldots, x_n\}$ be an object set, and $U = \{u_1, u_2, \ldots, u_m\}$ be a goal set. According to the method of Chang's extent analysis model, each object is taken and extent analysis for each goal, g_i, is performed respectively (Chang, 1992, 1996). Therefore, m extent analysis values for each object can be obtained with the following signs:

$$M_{gi}^1, M_{gi}^m, i = 1, 2, \ldots, n\ . \tag{4.3}$$

Where all the $M_{gi}^j (j = 1, 2, \ldots, m)$ are triangular fuzzy numbers. A triangular fuzzy number can be denoted as $M = (l, m, u)$ where $l \leq m \leq u$, l and u stand for the lower and upper value of the support of M, respectively, and m is the mid-value of M.

The steps of the improved Chang's extent analysis model, which is applied in this chapter, can be given as follows:

Step 1: The value of fuzzy synthetic extent with respect to the i th object is defined as:

$$S_i = \sum_{j=1}^{m} M_{gi}^j \otimes \left[\sum_{i=1}^{n} \sum_{j=1}^{m} M_{gi}^j \right]^{-1} . \tag{4.4}$$

To obtain $\sum_{j=1}^{m} M_{gi}^j$, perform the fuzzy addition operation of m extent analysis values for a particular matrix such that

$$\sum_{j=1}^{m} M_{gi}^j = \left(\sum_{j=1}^{m} l_j, \sum_{j=1}^{m} m_j, \sum_{j=1}^{m} u_j \right) \tag{4.5}$$

and to obtain $\left[\sum_{i=1}^{n} \sum_{j=1}^{m} M_{gi}^j \right]^{-1}$, perform the fuzzy addition operation of $M_{gi}^j (j = 1, 2, \ldots, m)$ values such that

$$\sum_{i=1}^{n} \sum_{j=1}^{m} M_{gi}^j = \left(\sum_{i=1}^{n} l_i, \sum_{i=1}^{n} m_i, \sum_{i=1}^{n} u_i \right) \tag{4.6}$$

and then compute the inverse of the vector in (4.6) such that

$$\left[\sum_{i=1}^{n}\sum_{j=1}^{m} M_{gi}^{j}\right]^{-1} = \left(\frac{1}{\sum_{i=1}^{n} u_i}, \frac{1}{\sum_{i=1}^{n} m_i}, \frac{1}{\sum_{i=1}^{n} l_i}\right). \quad (4.7)$$

The principles for the comparison of fuzzy numbers were introduced to derive the weight vectors of all elements for each level of the hierarchy with the use of fuzzy synthetic values. We now discuss these principles that allow the comparison of fuzzy numbers. (Zhu et al., 1999).

Step 2: The degree of possibility of $M_2 \geq M_1$ is defined as

$$V(M_2 \geq M_1) = \sup_{y \geq x} \left[\min\left(\mu_{M_1}(x), \mu_{M_2}(y)\right)\right], \quad (4.8)$$

where sup represents supremum (i.e., the least upper bound of a set) and when a pair (x, y) exists such that $y \geq x$ and $\mu M_1(x) = \mu M_2(y)$, then we have $V(M_2 \geq M_1) = 1$.

Since $M_1 = (l_1, m_1, u_1)$ and $M_2 = (l_2, m_2, u_2)$ are convex fuzzy number it follows that:

$$V(M_2 \geq M_1) = \text{hgt}(M_1 \cap M_2) = \mu M_2(d)$$

(where the term hgt is the height of fuzzy numbers on the intersection of M_1 and M_2)

$$\mu M2(d) = \begin{cases} 1, \text{if } m_2 \geq m_1 \\ 0, \text{if } l_1 \geq u_2 \\ \frac{l_1 - u_2}{(m_1 - u_2) - (m_1 - l_1)} \end{cases}. \quad (4.9)$$

Where d is the crossover point's abscissa of M_1 and M_2. To compare M_1 and M_2,we need both the values of $V(M_1 \geq M_2)$ and $V(M_2 \geq M_1)$.

Step 3: The degree of possibility for a convex fuzzy number to be greater than k convex fuzzy numbers M_i $(i = 1, 2, \ldots, k)$ can be defined by

$$\begin{aligned} &V(M \geq M_1, M_2, \ldots, M_k) \\ &= V[(M \geq M_1) \text{ and } M \geq M_2 \text{ and} \ldots \text{ and } (M \geq M_k)] \\ &= \min V(M \geq M_i), \quad i = 1, 2, 3, \ldots, k \end{aligned}$$

Assume that

$$d'(A_i) = \min \ V(S_i \geq S_k), \quad (4.10)$$

for $k = 1, 2, \ldots, n; k \neq i$. Then the weight vector is obtained as follows:

$$W' = (d'(A_1), d'(A_2), \ldots, d'(A_n))^T. \quad (4.11)$$

Where A_i $(i = 1, 2, \ldots, n)$ are n elements.

Step 4: After normalization, the normalized weight vectors are,

$$W = (d(A_1), d(A_2), \ldots, d'(A_n))^T. \quad (4.12)$$

Where W is not a fuzzy number.

4.4.1 A Numerical Application

As previously mentioned, in the third step of the framework, fuzzy AHP methodology is applied for weight determination. The sub-chains together with related risks are represented in Fig. 4.1.

In order to perform a pairwise comparison among the requirements, the linguistic scale as proposed by Büyüközkan (2008) and Büyüközkan et al. (2008) is adopted in this chapter. The scale is depicted in Fig. 4.2 and the corresponding explanations are provided in Table 4.4. Figure 4.2 shows the triangular fuzzy numbers $M = (l, m, u)$ where $l \leq m \leq u$, l and $u$stand for the lower and upper value of the support of M, respectively, and m is the mid-value of M. Similar to the importance scale defined in Saaty's classical AHP (Saaty, 1980), five main linguistic terms are used to compare the criteria: "equal importance (EI)", "moderate importance (MI)", "strong importance (SI)", "very strong importance (VSI)" and "demonstrated importance (DI)". Further, their reciprocals: "equal unimportance (EUI)", "moderate unimportance (MUI)", "strong unimportance (SUI)", "very strong unimportance (VSUI)" and "demonstrated unimportance (DUI)" have also been considered. For instance, if criterion A is evaluated "strongly important" than criterion B, then this answer means that criterion B is "strongly unimportant" than criterion A.

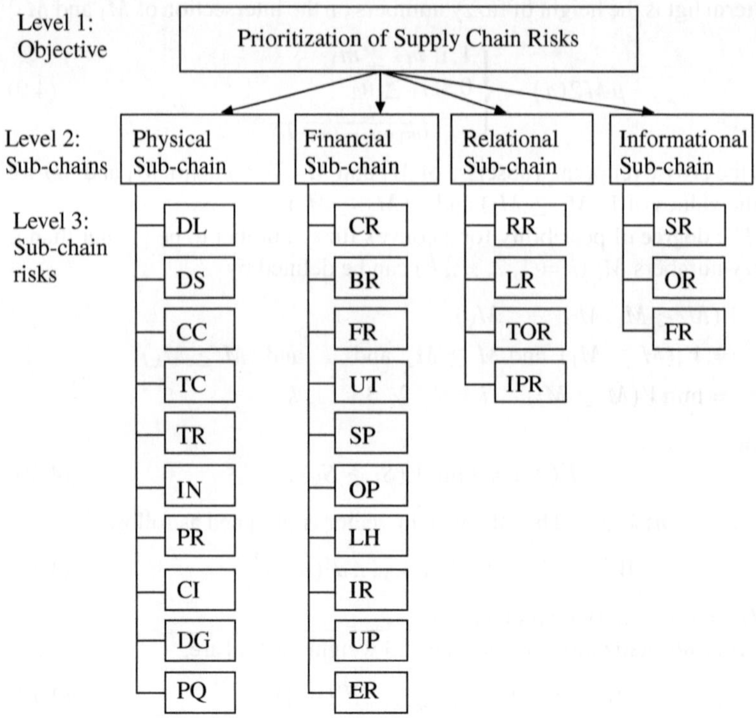

Fig. 4.1 A hierarchy based model of supply chain risks

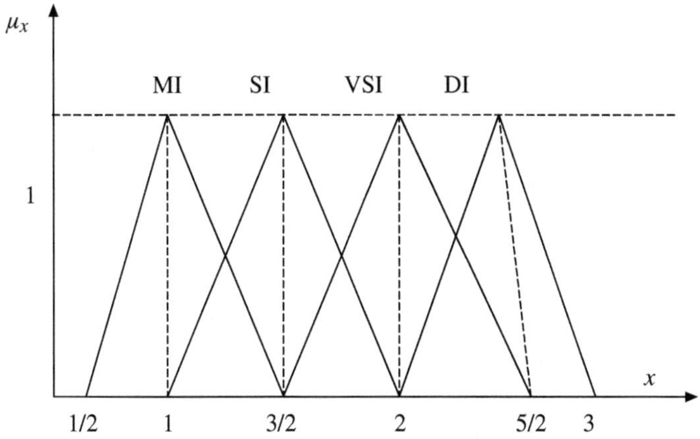

Fig. 4.2 Triangular fuzzy importance scale

Table 4.4 Triangular fuzzy importance scale

Linguistic Scale	Explanation	Triangular fuzzy Scale	Triangular fuzzy reciprocal scale
Equal Importance (EI)	Two requirements are the same importance	(1, 1, 1)	(1, 1, 1)
Moderate Importance (MI)	Experience and judgement slightly favor one requirement over another	(1/2, 1, 3/2)	(2/3, 1, 2)
Strong Importance (SI)	Experience and judgement strongly favor one	(1, 3/2, 2)	(1/2, 2/3, 1)
Very Strong Importance (VSI)	A requirement is favored very strongly over another; its dominance demonstrated in practice	(3/2, 2, 5/2)	(2/5, 1/2, 2/3)
Demonstrated Importance (DI)	The evidence favoring one requirement over another is the highest possible order of affirmation	(2, 5/2, 3)	(1/3, 2/5, 1/2)

Adapted from Büyüközkan (2008)

The proposed model was evaluated for small and medium enterprises (SMEs) cluster. A group of experts consisting of academics and professionals were asked to make pairwise comparisons for the sub-chains and their related risks mentioned in Sect. 4.5. A questionnaire (see Appendix A) is provided to get the evaluations. The overall results could be obtained by taking the geometric mean of individual

Table 4.5 Fuzzy evaluation matrix with respect to goal

	(PSC)	(FSC)	(RSC)	(ISC)
Physical sub-chain (PSC)	(1, 1, 1)	(1/2, 2/3, 1)	(1/2, 1, 3/2)	(3/2, 2, 5/2)
Financial sub-chain (FSC)	(1, 3/2, 2)	(1, 1, 1)	(1, 3/2, 2)	(3/2, 2, 5/2)
Relational sub-chain (RSC)	(2/3, 1, 2)	(1/2, 2/3, 1)	(1, 1, 1)	(1/2, 1, 3/2)
Informational sub-chain (ISC)	(2/5, 1/2, 2/3)	(2/5, 1/2, 2/3)	(2/3, 1, 2)	(1, 1, 1)

evaluations. However, since the group of experts came up with a consensus by the help of the Delphi Method in this case, a single evaluation could be obtained to represent the group's opinion (Bozbura et al., 2007; Büyüközkan, 2008) as represented in Table 4.5 for relative importance of sub-chain risks.

The values of fuzzy synthetic extents with respect to the sub-chain are calculated by applying formula (4.1) as below

$$\begin{aligned}
R_{\text{PSC}} &= (3.5,\ 4.66,\ 6) \otimes (0.0428,\ 0.0577,\ 0.0762) \\
&= (0.1498,\ 0.2688,\ 3.4572)\,, \\
R_{\text{FSC}} &= (4.5,\ 6,\ 7.5) \otimes (0.0428,\ 0.0577,\ 0.0762) \\
&= (0.1926,\ 0.3462,\ 0.5715)\,, \\
R_{\text{RSC}} &= (2.66,\ 3.66,\ 5.5) \otimes (0.0428,\ 0.0577,\ 0.0762) \\
&= (0.1138,\ 0.2111,\ 0.4191)\,, \\
R_{\text{ISC}} &= (2.46,\ 3,\ 4.33) \otimes (0.0428,\ 0.0577,\ 0.0762) \\
&= (0.1052,\ 0.1731,\ 0.3299)\,.
\end{aligned}$$

The degrees of possibility are calculated using these values and formula (4.5) as below:

$$\begin{aligned}
V(R_{\text{PSC}} \geq R_{\text{FSC}}) &= 0.1926 - 0.4572/(0.2688 - 0.4572) - (0.3462 - 0.1926) \\
&= 0.7736\,, \\
V(R_{\text{PSC}} \geq R_{\text{RSC}}) &= 1.00\,, \\
V(R_{\text{PSC}} \geq R_{\text{ISC}}) &= 1.00\,, \\
V(R_{\text{FSC}} \geq R_{\text{PSC}}) &= 1.00\,, \\
V(R_{\text{FSC}} \geq R_{\text{RSC}}) &= 1.00\,, \\
V(R_{\text{FSC}} \geq R_{\text{ISC}}) &= 1.00\,, \\
V(R_{\text{RSC}} \geq R_{\text{PSC}}) &= 0.1498 - 0.4191/(0.2111 - 0.4191) - (0.2688 - 0.1498) \\
&= 0.8235\,, \\
V(R_{\text{RSC}} \geq R_{\text{FSC}}) &= 0.1926 - 0.4191/(0.2111 - 0.4191) - (0.3462 - 0.1926) \\
&= 0.3289\,, \\
V(R_{\text{RSC}} \geq R_{\text{ISC}}) &= 1.00\,, \\
V(R_{\text{ISC}} \geq R_{\text{PSC}}) &= 0.1498 - 0.3299/(0.1731 - 0.3299) - (0.2688 - 0.1498) \\
&= 0.6530\,, \\
V(R_{\text{ISC}} \geq R_{\text{FSC}}) &= 0.1926 - 0.3299/(0.1731 - 0.3299) - (0.3462 - 0.1926) \\
&= 0.4423\,, \\
V(R_{\text{ISC}} \geq R_{\text{RSC}}) &= 0.1138 - 0.3299/(0.1731 - 0.3299) - (0.2111 - 0.1138) \\
&= 0.8504\,.
\end{aligned}$$

The weight vector of the main factors of the hierarchy can be calculated by using the formulas (4.10) and (4.6) as below:

$$
\begin{aligned}
d'(\text{PhysicalSC}) &= V(R_{\text{PSC}} \geq R_{\text{FSC}}, R_{\text{RSC}}, R_{\text{ISC}}) \\
&= \min(0.7736, 1, 1) = 0.7736\,, \\
d'(\text{FinancialSC}) &= V(R_{\text{FSC}} \geq R_{\text{PSC}}, R_{\text{RSC}}, R_{\text{ISC}}) \\
&= \min(1, 1, 1) = 1\,, \\
d'(\text{RelationalSC}) &= V(R_{\text{RSC}} \geq R_{\text{PSC}}, R_{\text{FSC}}, R_{\text{ISC}}) \\
&= \min(0.8235, 0.3289, 1) = 0.3289\,, \\
d'(\text{InformationalSC}) &= V(R_{\text{ISC}} \geq R_{\text{PSC}}, R_{\text{FSC}}, R_{\text{ISC}}) \\
&= \min(0.6530, 0.4423, 0.8504) = 0.4423\,, \\
W' &= (0.7736, 1, 0.3289, 0.4423)^{\text{T}}\,.
\end{aligned}
$$

Hence, via normalization, the normalized vectors of Physical, Financial, Relational, and Informational sub-chains risks are obtained as below:

$$
W_{\text{objective}} = (0.3039, 0.3929, 0.1292, 0.1738)^{T}
$$

In a similar way, the importance weights of the risks within physical sub-chain are calculated as follows

$$
\begin{aligned}
W =&(d(\text{DL}), d(\text{DS}), d(\text{CC}), d(\text{TC}), d(\text{TR}), d(\text{IN}), d(\text{PR}), d(\text{CI}), \\
&d(\text{DG}), d(\text{PQ}))^{\text{T}} \\
W_{\text{Physical}} =&(0.1123, 0.0679, 0.1231, 0.0697, 0.1298, 0.0753, 0.0526, 0.1336, \\
&0.0579, 0.1666)^{\text{T}}\,.
\end{aligned}
$$

It is observed that for the physical sub-chain poor quality, capacity inflexibility, transportation risks, product technological changes are more important than other risks.

In a similar way, the importance weights of the risks within financial sub-chain are calculated as follows

$$
\begin{aligned}
W =&(d(\text{CR}), d(\text{BR}), d(\text{FR}), d(\text{UT}), d(\text{SP}), d(\text{OP}), d(\text{LH}), d(\text{IR}), \\
&d(\text{UP}), d(\text{ER}))^{\text{T}}\,, \\
W_{\text{Financial}} =&(0.1121, 0.0893, 0.0547, 0.1443, 0.0582, 0.1204, 0.1318, 0.0357, \\
&0.1158, 0.1377)^{\text{T}}\,.
\end{aligned}
$$

It can be concluded that for financial sub-chain untimely payments, lack of hedging, and volatile oil prices emerge as the most important risks. In a similar way, the importance weights of the risks within relational sub-chain are calculated as follows

$$W = (d(\mathrm{RR}), d(\mathrm{TOR}), d(\mathrm{LR}), d(\mathrm{IPR}))^{\mathrm{T}},$$
$$W_{\mathrm{Relational}} = (0.2881, 0.2493, 0.2712, 0.1924)^{\mathrm{T}}.$$

For relational sub-chain, reputational risks and lack of trust and opportunism risk seem to appear more important than other risks. In a similar way, the importance weights of the risks within informational sub-chain are calculated as follows

$$W = (d(\mathrm{SR}), d(\mathrm{FR}), d(\mathrm{OR}))^{\mathrm{T}}$$
$$W_{\mathrm{Informational}} = (0.3799, 0.3761, 2440.)^{\mathrm{T}}.$$

It can be concluded that for informational sub-chain forecast risks emerges as the most important risk.

Finally, considering the obtained results, composite priority weights for supply chain risks can be calculated as given in Table 4.6.

Table 4.6 Composite priority weights for supply chain risks

Sub-chain	Local weights	Sub-chain risks	Local weights	Global weights
Physical	0.3039	DL	0.1123	0.0341
		DS	0.0679	0.0206
		CC	0.1231	0.0374
		TC	0.0697	0.0212
		TR	0.1298	0.0394
		IN	0.0753	0.0229
		PR	0.0526	0.0159
		CY	0.1336	0.0406
		DG	0.0579	0.0176
		PQ	0.1666	0.0506
Financial	0.3929	CR	0.1121	0.0440
		BR	0.0893	0.0351
		FR	0.0547	0.0215
		UT	0.1443	0.0567
		SP	0.0582	0.0229
		OP	0.1204	0.0473
		LH	0.1318	0.0518
		IR	0.0357	0.0140
		UP	0.1158	0.0455
		ER	0.1377	0.0541
Relational	0.1292	RR	0.2881	0.0372
		LR	0.2493	0.0322
		TOR	0.2712	0.0350
		IPR	0.1924	0.0248
Informational	0.1738	SR	0.3799	0.0660
		FR	0.3761	0.0653
		OR	0.2440	0.0424

Based on the values in Table 4.6 it can be concluded that forecast risks, system breakdown/security risks, untimely payment risks, exchange rate risks, lack of hedging risks, quality risks are the most important risks in supply chains as perceived by small and medium enterprises.

4.5 Concluding Remarks

A very important task in risk management is to establish those risk factors that are important to a particular company. With the help of this assessment the company is able to focus its resources more efficiently. The model presented in this chapter would help the practitioners to assign relative importance to various risks in a supply chain and develop plans accordingly to mitigate them.

Despite the recent surge in academic and practitioner publications regarding supply chain risks, the research presented in this chapter provides an additional value to the body of knowledge and consequently to managerial decision making. Because of the subjective and intangible nature of the risk variables considered in the model, the proposed methodology based on fuzzy-AHP framework provides a systematic method and is more capable of capturing a human's appraisal of ambiguity when complex multi-attribute decision-making problems are considered.

For further research, the model developed in this chapter can be evaluated for supply chains for large corporations. Also, other fuzzy multi-attribute approaches such as fuzzy TOPSIS and fuzzy outranking methods can be used for the prioritization of supply chain risks. In future models we can also consider the interdependence among various supply chain risks and in that case analytic network process (ANP) approach that takes into account the dependence and feedback can be applied to evaluate the model.

As risk is inherent in every link within a firm's supply chain it is impossible to completely insulate a supply chain from risks. But by understanding the sources of risk and prioritizing them, firms can take a proactive view for reducing and managing these risks.

Appendix

Sample questions from the questionnaire used to facilitate comparisons of sub-chain risks

Questionnaire

Read the following questions and put check marks on the pairwise comparison matrices. If an attribute on the left is more important than the one matching on the right, put your check mark to the left of the "Equal importance" column, under the importance level (column) you prefer. On the other hand, if an attribute on the left is less important than the one matching on the right, put your check mark to the right of the importance "Equal Importance" column, under the importance level (column) you prefer.

Questions

With respect to the overall goal "prioritization of the supply chain risks",

Q1. How important are physical sub-chain risks (PSR) when compared with financial sub-chain risks (FSR)?

Q2. How important are financial sub-chain risks (FSR) when compared with relational sub-chain risk (RSR)?

Q3. How important are informational sub-chain risks (ISR) when compared with financial sub-chain risks (FSR)?

Q4. How important are physical sub-chain risks (PSR) when compared with informational sub-chain risks (ISR)?

Q5. How important are physical sub-chain risks (PSR) when compared with relational sub-chain risks (RSR)?

Q6. How important are relational sub-chain risks (RSR) when compared with informational sub-chain risks (ISR)?

Questions	Sub-chain risks	Demonstrated Importance	Very Strong Importance	Strong Importance	Moderate Importance	Equal Importance	Moderate unimportance	Strong unimportance	Very Strong unimportance	Demonstrated unimportance	Sub-chain risks
1	PSR							√			FSR
2	FSR			√							RSR
3	ISR								√		FSR
4	PSR		√								ISR
5	PSR				√						RSR
6	RSR				√						ISR

References

1. Agrell P J, Lindroth R, Norrman A et al (2004) Risk, information and incentives in telecom supply chains. International Journal of Production Economics, 90:1–16
2. Barratt M (2004) Understanding the meaning of collaboration in the supply chain. Supply Chain Management: An International Journal, 9:30–42
3. Barratt M, Oliveira A (2001) Exploring the experiences of collaborative planning initiatives. International Journal of Physical Distribution & Logistics Management, 31:266–89
4. Bogataj D, Bogataj M (2007) Measuring the supply chain risk and vulnerability in frequency space. International Journal of Production Economics, 108:291–301
5. Bozbura F T, Beskese A (2007) Prioritization of organizational capital measurement indicators using fuzzy AHP. International Journal of Approximate Reasoning, 44:124–147
6. Bozbura F T, Beskese A, Kahraman C et al (2007) Prioritization of human capital measurement indicators using fuzzy AHP. Expert Systems with Applications, 32:1100–1112
7. Büyüközkan G (2008) Determining the mobile commerce user requirements using an analytic approach', Computer Standards & Interfaces, doi:10.1016/j.csi.2007.11.006
8. Büyüközkan G, Feyzioğlu O, Nebol E et al (2008) Selection of the strategic alliance partner in logistics value chain. International Journal of Production Economics, 113:148–158
9. Cakir O, Canbolat M S (2008) A web-based decision support system for multi-criteria inventory classification using fuzzy AHP methodology. Expert Systems with Applications, 35:1367–1378
10. Callioni G, Billington C (2001) Effective collaboration: Hewlett-Packard takes supply chain management to another level. OR/MS Today, 28:34–39
11. Cardinali R (1998) If the system fails, who is liable?. Logistics Information Management, 11:257–261
12. Cavinato J L (2004) Supply chain logistics risks. International Journal of Physical Distribution & Logistics Management, 34:383–387
13. Chan F T S, Kumar N (2007) Global supplier development considering risk factors using fuzzy extended AHP-based approach. Omega, 35:417–431
14. Chang D-Y (1992) Extent Analysis and Synthetic Decision, Optimization Techniques and Applications. World Scientific, Singapore
15. Chang D-Y (1996) Applications of the extent analysis method on fuzzy AHP. European Journal of Operational Research, 95:649–655
16. Chopra S, Sodhi M S (2004) Managing risk to avoid supply chain breakdown. Sloan Management Review, 46:53–61
17. Christopher M, Lee H (2004) Mitigating supply chain risk through improved confidence. International Journal of Physical Distribution & Logistics Management, 34:388–396
18. Christopher M, McKinnon A, Sharp J, Wilding R, Peck H, Chapman P, Jüttner U, Bolumole Y (2002) Supply Chain Vulnerability, Cranfield University, Cranfield
19. Craighead C W, Blackhurst J, Rungtusanatham M J and Handfield R B (2007) The Severity of Supply Chain Disruptions: Design Characteristics and Mitigation Capabilities. Decision Sciences, 38:131–156
20. Cucchiella F, Gastaldi M (2006) Risk management in supply chain: a real option approach. Journal of Manufacturing Technology Management, 17:700–720
21. Durán O, Aguilo J (2008) Computer-aided machine-tool selection based on a Fuzzy-AHP approach. Expert Systems with Applications, 34:1787–1794
22. Elkins D, Handfield R B, Blackhurst J, Craighead, C W et al (2005) 18 ways to guard against disruption. Supply Chain Management Review, 46–53
23. Erensal Y C, Oncan T, Demircan M L (2006) Determining key capabilities in technology management using fuzzy analytic hierarchy process: A case study of Turkey. Information Sciences. 176:2755–2770
24. Faisal M N, Banwet D K, Shankar R et al (2006) Mapping supply chains on risk and customer sensitivity dimensions. Industrial Management & Data Systems,106:878–895

25. Faisal M N, Banwet D K, Shankar R et al. (2007a) Supply chain risk management in SMEs: analysing the barriers. International Journal of Management and Enterprise Development, 4:588–607
26. Faisal M N, Banwet D K, Shankar R et al. (2007b) Information risk management in supply chains: an assessment and mitigation framework. Journal of Enterprise Information Management, 20:677–699
27. Finch P (2004) Supply chain risk management. Supply Chain Management: An International Journal, 9:183–196
28. Frenkel S J, Scott D (2002) Compliance, collaboration, and codes of labor practice: the Adidas connection. California Management Review, 45:29–49
29. Fu H-P, Ho Y-C, Chen R.C.Y, Chang T-H, Chien P-H et al (2006) Factors affecting the adoption of electronic marketplaces. International Journal of Operations & Production Management, 26:1301–1324
30. Giunipero L C, Pearcy D H (2000) World-class purchasing skills: an empirical investigation. Journal of Supply Chain Management, 36:4–13
31. Giunipero L C, Eltantawy R A (2004) Securing the upstream supply chain: a risk management approach. International Journal of Physical Distribution & Logistics Management, 34:698–713
32. Goh M, Lim J Y S, Meng F et al (2007) A stochastic model for risk management in global supply chain networks. European Journal of Operational Research,182:164–173
33. Gunasekaran G, Patel C, McGaughey R E (2004) A framework for supply chain performance measurement. International Journal of Production Economics, 87:333–347
34. Hale T, Moberg C R (2005) Improving supply chain disaster preparedness. International Journal of Physical Distribution & Logistics Management, 35:195–207
35. Hallikas J, Karvonen I, Pulkkinen U, Virolainen V M, Tuominen M et al (2004) Risk management processes in supplier networks. International Journal of Production Economics, 90:47–58
36. Hallikas J, Virolainen V M, Tuominen M (2002) Risk analysis and assessment in network environments: A dyadic case study. International Journal of Production Economics, 78:45–55
37. Harland C, Brenchley R, Walker H et al (2003) Risk in supply networks. Journal of Purchasing and Supply Management, 9:51–62
38. Hauser L M (2003) Risk-adjusted supply chain management. Supply Chain Management Review, Nov/Dec:64–71
39. Hendricks K B, Singhal V R (2003) The effect of supply chain glitches on shareholder wealth. Journal of Operations Management, 21:501–522
40. Hendricks K B, Singhal V R (2005) An Empirical Analysis of the Effect of Supply Chain Disruptions on Long-Run Stock Price Performance and Equity Risk of the Firm. Production and Operations Management, 14:35–52
41. Jiang B, Baker R C, Frazier G V et al (2007) An analysis of job dissatisfaction and turnover to reduce global supply chain risk: Evidence from China. Journal of Operations Management, doi:10.1016/j.jom.2007.09.002
42. Johnson M E (2001) Learning from toys: Lessons in managing risk from toy industry. California Management Review, 43:106–125
43. Jüttner U (2005) Supply chain risk management: Understanding the business requirements from a practitioner perspective. The International Journal of Logistics Management, 16:120–141
44. Jüttner U, Peck H, Christopher M et al (2002) Supply chain risk management: outlining an agenda for future research, Griffiths, J., Hewitt, F., Ireland, P., Proceedings of the Logistics Research Network 7th Annual Conference, 443–450
45. Kang H-E, Lee A H I (2007) Priority mix planning for semiconductor fabrication by fuzzy AHP ranking Expert Systems with Applications. 32:560–570
46. Kaufmann A, Gupta M M (1991) Introduction to Fuzzy Arithmetic Theory and Applications. Van Nostrand Reinhold, New York

47. Kern T, Willcocks L (2000) Exploring information technology outsourcing relationships: theory and practice. The Journal of Strategic Information Systems, 9:321–350
48. Kleindorfer P R, Saad G H (2005) Managing Disruption Risks in Supply Chains. Production and Operations Management, 14:53–68
49. Laarhoeven P J M, Pedrycz W (1983) A fuzzy extension of Saaty's priority theory. Fuzzy Sets and Systems, 11:229–241
50. McFadden P J (1997) Guarding computer data. Journal of Accountancy, 184:77–79
51. Mentzer J T, Foggin J H, Golicic S G (2000) Supply chain collaboration: enablers, impediments, and benefits. Supply Chain Management Review, 4:52–58
52. Neiger D, Rotaru K, Churilov L. (2008) Supply chain risk identification with value-focused process engineering. Journal of Operations Management, doi:10.1016/j.jom.2007.11.003
53. Norrman A, Jansson U (2004) Ericsson's proactive supply chain risk management approach after a serious sub-supplier accident. International Journal of Physical Distribution & Logistics Management, 34:434–456
54. Ojala M, Hallikas J (2006) Investment decision-making in supplier networks: Management of risk. International Journal of Production Economics, 104:201–213
55. Peck H (2005) Drivers of supply chain vulnerability: an integrated framework. International Journal of Physical Distribution & Logistics Management, 35:210–232
56. Piplani R, Fu Y (2005) A coordination framework for supply chain inventory alignment. Journal of Manufacturing Technology Management, 16:598–614
57. Ritchie B, Brindley C (2000) Disintermediation, disintegration and risk in the SME global supply chain. Management Decision, 38:575–583
58. Saaty T L (1980) The Analytic Hierarchy Process, McGraw Hill, New York
59. Sarkar A, Mohapatra P K J (2006) Evaluation of supplier capability and performance: A method for supply base reduction. Journal of Purchasing & Supply Management, 12:148–163
60. Sasmal S, Ramanjaneyulu K (2008) Condition evaluation of existing reinforced concrete bridges using fuzzy based analytic hierarchy approach. Expert Systems with Applications, 35:1430–1443
61. Shi D (2004) A review of enterprise supply chain risk management. Journal of Systems Science and Systems Engineering, 13:219–244
62. Sinha P R, Whitman L E, Malzahn D et al (2004) Methodology to mitigate supplier risk in an aerospace supply chain. Supply Chain Management: An International Journal, 9:154–168
63. Sounderpandian J, Prasad S, Madan M et al (2008) Supplies from developing countries: Optimal order quantities under loss risks. Omega, 36:122–130
64. Souter G (2000) Risks from supply chain also demand attention. Business Insurance, 34:26–28
65. Speckman R E, Davis E W (2004) Risky Business: expanding the discussion on risk and the extended enterprise. International Journal of Physical Distribution & Logistics management, 34:414–433
66. Svensson G (2004) Key areas, causes and contingency planning of corporate vulnerability in supply chains: A qualitative approach. International Journal of Physical Distribution & Logistics Management, 34:728–748
67. Tang C S (2006) Robust Strategies for Mitigating Supply Chain Disruptions. International Journal of Logistics: Research and Applications, 9:33–45
68. Treleven M, Schweikhart S B (1988) A risk/benefit analysis of sourcing strategies: Single vs multiple sourcing. Journal of Operations Management, 7:93–114
69. Tsai C-Y (2008) On supply chain cash flow risks. Decision Support Systems, 44:1031–1042
70. Yu C S (2002) A GP-AHP method for solving group decision-making fuzzy AHP problems. Computers and Operations Research, 29:1969–2001
71. Zadek S (2004) The path to corporate responsibility. Harvard Business Review, 82:125–132
72. Zhu K-J, Jing Y, Chang D Y (1999) A discussion on extent analysis method and applications of fuzzy AHP. European Journal of Operational Research, 116:450–456

73. Zsidisin G A, Panelli A, Upton R et al (2000) Purchasing organization involvement in risk assessments, contingency plans, and risk management: an exploratory study, Supply Chain Management: An International Journal, 5:187–198
74. Zsidisin G A, Ellram L M, Carter J R, Cavinato J L et al (2004) An analysis of supply risk assessment techniques. International Journal of Physical Distribution & Logistics Management, 34:397–413

Chapter 5
A Generalized Simulation Framework for Responsive Supply Network Management

Jin Dong, Wei Wang, and Teresa Wu

Abstract Firms are under the pressure to explore various strategies to improve the supply network performance so that customers' demands can be met more responsively. Many of the challenges from implementing the strategies lie in the distributed and dynamic nature of the network where geographically dispersed entities may have different goals and objectives. Additionally, irregularities and disruptions occurring at any point in the network may propagate through the network and amplify the negative impact. These disruptions, often occurring without warning due to the dynamic nature of a supply network, can lead to poor performance of the supply network. A key component in responsive supply network management is to proactively assess the robustness and resilience to disruption of a supply network. Discrete Event Simulation (DES) can achieve this. In this chapter, we introduce a simulation tool developed by IBM China Research Lab, named General Business Simulation Environment (GBSE). It can capture supply network dynamics with a fine level of granularity and provide useful insights to supply network's real operations. GBSE is designed for tactical-level decision making, and may be useful for supply network what-if analysis and risk analysis. The architecture of GBSE is detailed in this chapter followed by several scenarios in an automobile supply network to demonstrate the applicability of GBSE to assess the responsiveness of a supply network.

5.1 Introduction

Supply network management focuses on integrating material flows and information flows to increase the value and responsiveness for the customer. However, such integration is not an easy task due to the diversity of the participants in terms of size, technological capabilities, culture differences and efficiencies (Blackhurst et al., 2004). One additional difficulty is the risk inherent in the network, which refers to the potential deviations from the original objectives. This can cause decreases in value added activities at different levels in the netowrk. Therefore, to have a complex supply network with stable performance, risk assessment is an integral part of the management process. A large number of analytical models (deterministic optimization, stochastic optimization) have been developed to study supply network for profit maximization and risk minimization. However, these models suffer from some shortfalls that limit their applications to supply network risk management: (1) the size and complexity of a typical network means that the mathematical models can involve a very large number of variables and constraints. Maintaining such models can be difficult while the computational burden can be heavy. (2) Many assumptions, such as the linearity assumption, to simplify the model may not hold (Wu and O'Grady, 2004). On the other hand, discrete event simulation (DES) has been proven valuable to be a practical tool for representing complex interdependencies, evaluating alternative designs and policies, and analyzing performance tradeoffs for supply chain systems (Hennessee, 1998; Chwif et al., 2002; Jain et al., 2002; Venkateswaran et al., 2002; Enns and Suwanruji, 2003; Gan et al., 2000).

However, simulation is not without limitations. First, it is a challenge for the analyst to determine the proper level of granularity. Secondly, many skills are required to create a simulation model, and sometimes it is necessary to write programming codes for special scenarios. Thirdly, a large volume data is required to develop realistic simulation models. Lastly, it is time-consuming to run some simulations, for instance, it takes several hours to run one medium-size scenario (Dong et al. (2006)). To address these issues, IBM China Research Lab developed a supply chain simulation tool, named General Business Simulation Environment (GBSE), which is a flexible and powerful software tool to help supply chain practitioners to model, simulate and analyze their supply networks. GBSE is previously a part of IBM SmartSCOR (Dong et al. (2006)), and it is usually used for making tactic-level decisions. However, it does have the ability for developing interfaces, linking external tools for strategic and/or operational level models. GBSE has been applied to study various supply networks such as wholesale industry and automobile industry, just to name a few. In this study, of particular interest is the application of GBSE to assess the robustness of a supply network, specifically, an automobile supply network. We first review supply chain simulation tools and framework followed by detailed description of GBSE architecture. We then explain the development of the supply chain simulation model developed using GBSE. Four scenarios are studied to assess the supply chain performance with respect to varied demand forecasts, varied selections of supplier, shipment and production.

5.2 Review of Supply Chain Simulation

Simulation has been commonly used tool to supply network management and several comprehensive surveys have been conducted to summarize the applications of simulation. For instance, Kleijnen (2003) provides a survey of simulation in supply chain management and categorizes the simulation into four types: spreadsheet simulation, system dynamics, discrete event simulation (DES) and business games.

Terzi and Cavalieri (2004) focus on the architecture of the simulation and conduct comparison studies between local monolithic simulation, parallel and distributed simulation paradigms. Compared to local simulation, distributed simulation can leverage computation resources from multiple machines to handle larger problems within a shorter time. However, in supply chain simulation area, local simulation is still popular because of its simplicity. In local simulation, Monte Carlo Simulation (MCS) and DES are two important methods. MCS is a light-weight static method which can be implemented with spreadsheet based tools. It is very useful for creating high-level models for preliminary results. DES is more computational intensive methodology with the capability to handle the modeling and decisions in great details.

Applicability of simulation for supply network decision-making has been well studied. For instance, Lendermann et al. (2001) demonstrate the use of simulation in semiconductor manufacturing. Its popularity is also reflected in industry applications. As early as in late last century, IBM developed a supply chain simulator, which has a mix of simulation and optimization functions to model and analyze IBM supply chains (Bagchi et al., 1998). In 1999, IBM employed its own simulation-based supply chain analyzer to visualize and quantify the effects of making changes on a hypothetical supply chain, and the impact of the changes on system performance (Archibald et al., 1999). Further, the need for executing supply chain simulations based on a full-detailed model has also been pointed out: Jain et al. (1999) compare two models with different levels of detail for semiconductor manufacturing supply chains and concluded that simulations incorporating detailed models are required when attempting to determine the correct inventory levels for maintaining desired customer responsiveness. In these studies, abstracted models may result in inaccurate solutions that subsequently lead to erroneous decisions. Similar conclusions are draw by Venkateswaran et al. (2002). Thus, it is necessary to develop a simulation tool with fine granularity which enables accurate assessment of the supply network.

There exist a number of general purpose simulation tools for supply network modeling including Arena, AnyLogic, AutoMod, just to name a few. When using these tools, analysts usually create supply chain models with a set of pre-defined logic elements, like CreateNode, DecisionNode, DisposeNode, etc. Most existing tools have separate views for logic and presentation where logic view presents the internal simulation logic with flow charts, decision trees, and presentation view shows animations, charts, maps. While promising, the importance of the supply network drives the effort to develop specific packages to simulate supply chain. Swaminathan et al. (1998) propose a supply chain modeling framework, in which supply chain models are composed from software components that represent types of supply chain agents, their constituent control elements, and their interaction protocols. Rossetti and Chan (2003) discuss the design, development and testing of an object-oriented prototype framework for supply chain simulation. Later on, Rossetti et al. (2006) develop a JSL (Java Simulation Library) based object-oriented framework for simulating multi-echelon inventory systems. EasySC is a simulation platform for understanding supply chains through studying the impact of stochastic demands, logistics decisions and production policies on key performance measures

(Liu et al., 2004). Supply Chain Guru (Supply Chain Guru, 2008) is a supply chain simulation and optimization tool for enterprise strategic planning and targeted supply chain performance improvements. Other than these full scale frameworks and systems reviewed above, numerous simulation libraries are available to facilitate building simulation tools. Repast (Repast, 2008) and Scalable Simulation Framework (SSF, 2008) is one such example. As a leading firm in studying supply network perfoamance, IBM China Research Lab recently developed a large scale Discrete Event Simulation package, named GBSE which is extensible and flexible with user friendly interfaces. This package has been applied to provide consulting solutions to various industries ranging from wholesale industry, financial services to automobile industry. In the following section, the framework and architecture of GBSE is explained in details.

5.3 GBSE: A Supply Chain Simulation Environment

Figure 5.1 illustrates a typical supply chain system modeled by GBSE. The supply chain operation consists of two processes: planning and execution. The *planning process* includes demand forecasting which is mainly based on historical order data and supply planning which creates a supply commitment from components suppliers. The *execution process* starts at the order processing once the orders are received from customers. The order fulfillment and scheduling module generates the pur-

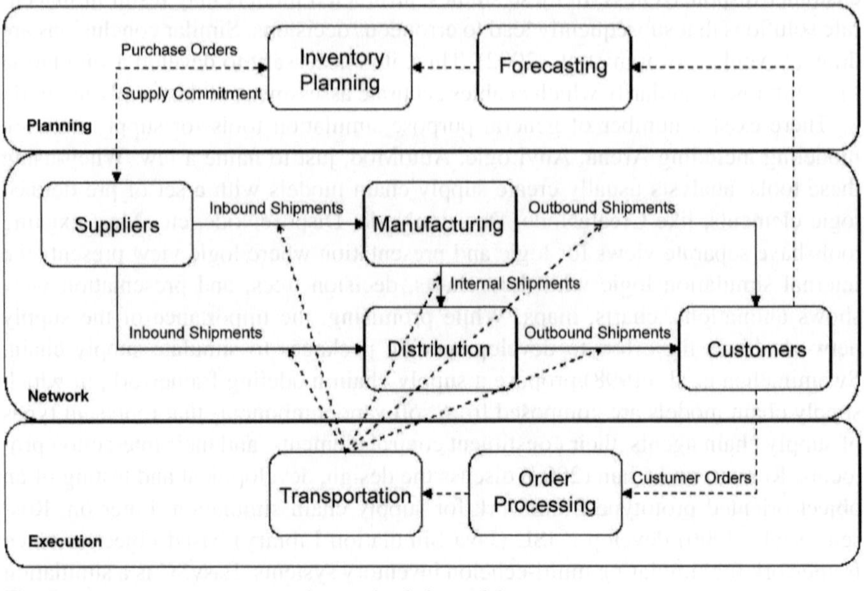

Fig. 5.1 Functional components of a supply chain model

chase orders to the suppliers as well as the premium shipping approvals which authorize the use of shipment modes (air vs. ocean in this study). The *network* process models a general supply network consisting of supplier, manufacturing, distribution and customers.

The simulation engine models six important components/functions of a supply chain network including:

- **Customer**: representing external customers that issue orders to the supply chain, based on demand forecast. GBSE also allows the user to set the desired service level and priority for the customer.
- **Manufacturing**: modeling assembly process and keeping raw materials and finished goods inventory.
- **Distribution**: modeling distribution centers, including finished goods inventory and material handling.
- **Transportation**: modeling transportation time, vehicle loading, and transportation costs.
- **Forecasting**: modeling demand forecast of products, including promotional and stochastic demand.
- **Inventory planning**: modeling periodic setting of inventory target levels. Underlying this process is the GSBE optimization engine that computes recommended inventory levels at various locations in the supply chain based on desired customer serviceability.

In addition, the simulation engine provides an animation module that allows users to see how materials and information flow through the supply chain.

5.3.1 GBSE Architecture

As a flexible framework that can be extended to be compatible with other existing system, GBSE is designed to have layered architecture (Fig. 5.2): data layer, controller layer, service layer, and presentation layer. The Data Layer manages all the data in simulation; the Service Layer defines all the simulation processes; the Con-

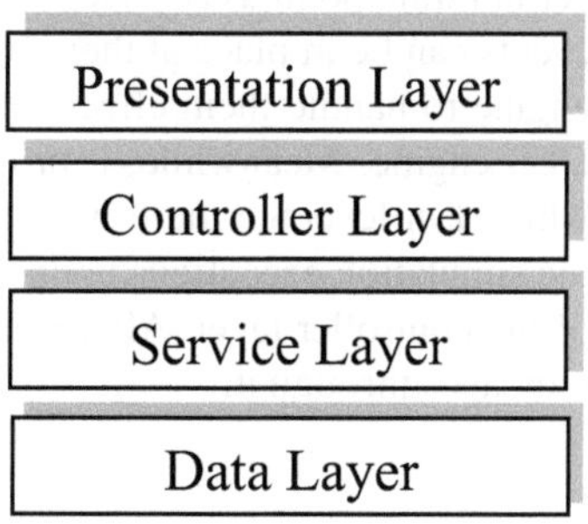

Fig. 5.2 GBSE architecture

troller Layer implements the simulation engine; and the Presentation Layer is the interface to simulation analyst.

5.3.1.1 Data Layer

The Data layer manages all the simulation data, including configuration data, run-time data and report data. A non-trivial simulation model in general involves large volume data. It is time-consuming to access and handle all the data, which prohibits the detailed study of a supply network. Thus, an efficient data access solution is important to the success of simulation study. In GBSE, the data access layer is built upon a relational database system which caches the data to manage the out-of-memory issue. Data can be imported and exported to files with different formats, for instance, Microsoft Excel files and CSV (Comma Separated Version) files.

5.3.1.2 Service Layer

The Service layer contains simulation services which are the finest unit in a simulation execution. Four types of data are associated with a service: configuration data, input data, output data, and state data. Each service has its internal transactions to operate on inputs and generate outputs. The behavior of services is controlled by the configuration data with the runtime status being stored in state data. Note that the Data layer is leveraged by Service layer to manage these four types of data. Simulation logic is implemented as one service or a combination of several services. Services can be grouped as service bundles.

5.3.1.3 Controller Layer

The Controller layer is responsible for scheduling service events and dispatching messages between services. A discrete event simulation (DES) engine and a simulation bus are implemented in this layer. The DES engine is the heart of the controller layer. An event list is maintained in the engine, and events are scheduled to occur at certain time point as services request. In a non-trivial simulation study, millions of events can be in place at the same time, so it could be a big challenge for the DES engine to handle them effectively. GBSE implements a built-in high-performance DES engine. Meanwhile, it provides the interface to other external DES engines which can leverage the usages between internal and external engines to optimize the simulation executions upon design request. The simulation bus is the backbone of the controller layer. All services are connected to the service bus and exchange messages through it.

5.3.1.4 Presentation Layer

The Presentation layer is the interface to the end users. It helps users to build the model, run the simulation, and perform analysis with reports. There are two perspectives: the modeling perspective (Fig. 5.3a) is used to create simulation model, and the running perspective (Fig. 5.3b) is used to launch the simulation and monitor the process. In addition, we have implemented five functions in the presentation layer: (1) create and edit the simulation model; (2) import and export the simulation data; (3) control the simulation running; (4) monitor the simulation running status; (5) generate and check the simulation report.

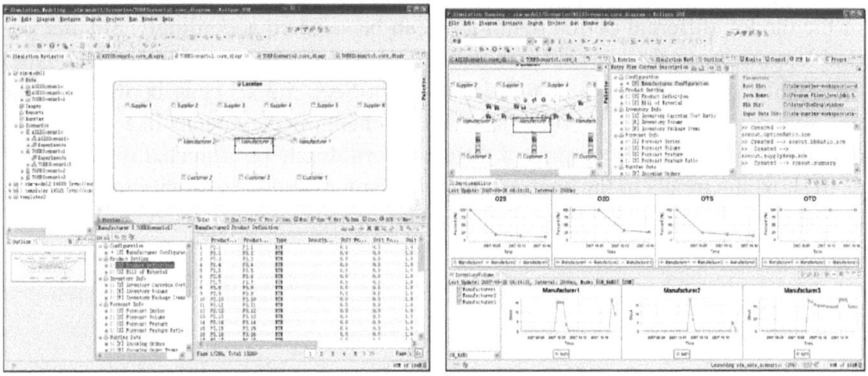

a b

Fig. 5.3 **a** Modeling perspective. **b** Running perspective

5.3.2 *GBSE for Supply Chain Simulation*

The four layers explained above are leveraged to create simulation models through the use of templates. A template is a module which consists of (1) simulation elements and (2) customized user interfaces. Each simulation element consists of data definitions and services. Services can be reused in several simulation elements. As an example, in GBSE we can have a template for a distribution network, and another template for a banking system. In the distribution network template, simulation elements are distribution centers, warehouses, hubs, retails, routes, etc. In the banking system template, simulation elements are banking branches, ATMs, customers, etc. GBSE provides different palettes for different templates. After applying a template to a model, analysts can drag elements from the palette and drop them to the model. A uniform table editor is provided for editing data associated with each element. The particular interest in this study is the Supply Chain Simulation Template. It implements supply chain model at tactical level, and allows analysts to create a multi-period and multi-echelon supply chain model. The data structure, modeling processes and performance measurement for a supply network are described.

5.3.2.1 Data Structure

All data is managed in data layer and stored in database. Collecting and cleaning up data is the one of the most time-consuming part of simulation modeling, and the success of a simulation study depends on the quality of data directly. In GBSE, four types of data are considered: Calendar, Product, Network and Configuration.

Calendar Data

GBSE adopts a multi-period simulation model. A period can be modeled as a day, a week, a month, a quarter, or even a year. Depending on the expected granularity of the model, different period definitions can be selected accordingly. Smaller period lengths mean finer granularity and thus more requirements on the data. Note the simulation configuration data (discussed in Sect. 5.3.2.1) can be different for different periods. This is useful to model time-changing parameters like seasonality demand. The simulation outputs of different periods can be collected separately so as to show the trends to the analysts. Supply chain planning is heavily relied on the calendar data to define the planning horizons.

Product Data

The Product data represents the physical components flowing through the network including finished goods, raw materials and intermediate products. It is not practical to have the information of all the components in the network recorded, so analysts need to select a proper subset since most configuration data is specific to certain kinds of products. General information for products should be provided, such as weight, dimension and brand. If manufacturing processes are taken into account, analysts will need to determine bill of material (BOM), which defines the composition relationships between different levels of products. Again, analysts should make the selection of the parts based on certain rules, such as selecting most expensive or heaviest parts.

Network Data

A supply chain network consists of nodes and links. In GBSE simulation model, we define three types of nodes (Customer, Supplier, and Facility) and one type of link (Lane). Supplier nodes are the source of supply, and they are the origins of a supply chain. Since there are no upstream nodes for suppliers, we assume the inventories are replenished by rules. In addition, supply capacity is taken into consideration, that is, the supply volume within specified periods can not exceed the allowed capability. Customer nodes are the source of demand, and they are final destinations of a supply chain. Depending on the size of the supply chain, customers can be clustered to generate the demands at various levels ranging from a country to a shop. Facility nodes are internal sites including warehouses, distribution centers, factories, retailers, wholesalers, etc. In GBSE, facilities are modeled with more details than suppliers and customers to assist the study. Lanes are physical directed connections between nodes. Inbound lanes connect suppliers and facilities; outbound lanes connect facilities and customers; inter-facility lanes connect facilities and facilities. There should be no more than one lane between two nodes, but there can be multiple transportation modes associated with one lane. If geographical information is provided, such as latitude and longitude, the supply chain network can be displayed in a geographic information system (GIS) embedded in GBSE. This gives analysts an intuitive view of the entire network.

Configuration Data

Configuration data refers to demand data, cost data, capacity data, lead time data, policy data. Demand data is defined for Customer Nodes. Usually only aggregated demand are available. GBSE can split the aggregated demand into a number of orders. Cost data includes one-time transition cost, fixed cost and variable cost. One-time transition cost is considered only under the situation that some existing facilities are closed or some potential facilities are opened. Fixed cost applies for specific products (defined by the analyst) while variable cost depends on product volume. In a multiple period model, both fixed cost and variable cost are related to periods, that is, for different periods, the costs can be different. Capacity data consists of supply capacity, manufacturing capacity, storage capacity, handling capacity, transportation capacity. Lead time data represents manufacturing lead time, transportation lead time, handling lead time, etc. Usually lead time is defined as a random number. Policy data is parameters to different policies, for instance, the batch size in transportation policy. There also exists other data such as customer order interval, shipping shrinkage, manufacturing yield, budget, time threshold in GBSE.

5.3.2.2 Supply Chain Modeling Processes

A process is a working unit at operational level which has its inputs and outputs, as well as internal logic to handle inputs and generate outputs. Meanwhile, processes can be controlled by local parameters and global parameters. In GBSE, all processes are implemented as services in service layer.

Resource Management

In supply chain context, resources can be workers, trucks, drivers, and handling equipment with limited capacities. GBSE implements a generic resource management framework in which the key components are resources in a resource pool. A simulation task can request a resource from the pool, and return the resource to the pool when the task is terminated. There are also cases that tasks can consume specific resources, as a result, the resource pool will manage the replenishment.

Customer Order Generating Process

The Customer Order Generating Process applies to Customer Node. Usually, aggregated demand data is available for the customers. In the simulation, the aggregated demand will be split into several waves. In each wave, one or more customer orders are generated and sent to corresponding facilities. Forecast errors are applied to order quantity which reflects the demand uncertainty. Since the demand is split into a few waves, the forecast error can be applied to either the total demand of the period, or the order quantity in one wave.

Order Process

Once an order enters a facility, it will be first pre-processed followed by putting into a prioritized order queue. The order queue will be checked periodically. For each round, some orders will be released and leave the order queue to get processed. Orders with higher priority will leave the queue earlier than orders with lower priority. An availability check is performed for each order to determine whether it can be released. In GBSE, the on-hand stock is checked first. If the order can not be fulfilled at the on-hand stock level, the in-transit stock will be checked. At this step, the due time needs to be estimated and compared with the arrival time of in-transit shipments. In the cases that in-transit stock can not fulfill the order, special arrangements have to be made such as placing urgent purchase orders. If still unsuccessful, the order will stay in the order queue and waiting for the next round.

Inventory Process

Inventory is considered as the control center as it triggers the events such as procurement when the inventory level is low than required. As a matter of fact, products fall into two categories: in-source and out-source. For in-source products, the replenishment is built from the manufacturing process in the same facility; for out-source products, the replenishment is from external suppliers via procurement process. Inventory control is applied to each product in each facility. Analysts can specify periodical review or continuous review. Several typical inventory control policies are implemented including (R, Q) and (s, S). These parameters can be set for each product and each period. If required, handling process can be modeled, which represents the tasks to move products into the storage area or move them out. This will involve resource framework to manage the handling equipments.

Procurement Process

The Procurement process is in charge of purchasing products from upstream facilities or suppliers. We model two procurement policies: single sourcing and multiple sourcing. In single sourcing policy, the analyst needs to locate the supplier for each product in each period. As for multiple sourcing policy, a set of methods are defined in GBSE for supplier selection. As an example, analysts can define the volume proportion for each supplier for each product. Procurement process can be triggered by procurement plan and inventory signals. Procurement plan defines the time and quantity of each procurement task, which allows analysts the complete control of procurement process. Meanwhile, inventory control process can trigger procurement tasks for out-source products when the inventory level is lower than a predefined reorder point. GBSE consolidates the procurement tasks by grouping the orders to the same suppliers based on the desired delivery date.

Manufacturing Process

GBSE focuses on discrete-manufacturing. An assembly process is implemented to model the consumption of raw materials and create intermediate and finished goods. Similar to the procurement process, the manufacturing process can be triggered by manufacturing plan and inventory signals. A Manufacturing plan consists of a set of manufacturing tasks; each task has its time and quantity. An Inventory signal is sent from inventory process for in-source products. Manufacturing capacities are defined to restrict the maximum manufacturing volume for specified products and periods. On the other hand, analysts are allowed to define manufacturing resources such as machines, workers, etc. Note that the manufacturing process will halt in the circumstances that corresponding resources are not available. GBSE will calculate the overall manufacturing cost consisting of direct labor cost, in-direct labor cost, energy consumption, and all other variable costs during manufacturing.

Transportation Process

The Transportation process happens on inbound, outbound, and inter-facility lanes. Several types of transportation modes are supported in GBSE, like LTL (Less-than-Truckload), TL (Truckload), ocean, train, and air shipping, other transportation modes, if not available, can be added by the analysts. Transportation time, cost, as well as the desired service level, are defined for each mode. GBSE can consolidate the transportation tasks which can save the transportation expenses. These consolidated tasks are grouped based on customers and due time, and the fulfilled orders are sent to customers in batches.

Planning Process

To enable BTP (Build-to-Plan) process, it is necessary to run MRP process during the simulation. Several parameters and system status will be taken into account by MRP process (1) for demand, average customer demands in the next several periods should be provided; (2) for inventory, backlog, in-transit inventory, and on-hand inventory should be calculated together; (3) for time, supply lead time and delivery lead time should be estimated. With all the three kinds of data as input, procurement quantity and procurement time can be determined. A planning engine is implemented in GBSE which provides the interfaces to external engines.

5.3.2.3 Performance Measurement

One important target of GBSE is to balance service level and total cost. Service Level is measured by time and percentage. Time related metrics show supply chain responsiveness. Some important metrics may be of interest to analysts are order cycle time, average assembling time, average transportation time, etc. It is also useful to check the average waiting time for an order queue or a resource pool. Percentage related metrics represent supply chain effectiveness. Some metrics of interest are the order fill rates, procurement fulfill rates, out-of-stock rates, etc. GBSE can record the related metric to enable the study of the supply network performance quantitatively. Due to the fact that (1) GBSE considers uncertain inputs (2) various what-if setting can be easily modeled in GBSE, in this chapter, we assess the supply network performance under different demand forecast, different supplier selection, different shipment selection and different production selection. This will provide the insights to design a more responsive supply network.

5.4 Experiments

In this section, we present a case study for an automotive manufacturer with GBSE. Simulation has been applied at three different levels: production line level, individual plant level, and supply chain level. In production line level, simulation can be used for product design; analysts create continuous model for physical systems and evaluate the performances. In plant level, simulation can be used to design the layout of the shop floor. This study will be focused on supply chain level. Current practices of automotive industry indicate the main challenges are:

- The Bill-of-Material (BOM) is very complex: 1) The number of components is usually quite large: thousands of components are required to produce a car. 2) A substitute BOM is very common in the automotive industry. Some components can be replaced by other components. 3) Components have a life cycle. Some components will not be used any more after the specified time point.

4) During the manufacturing process, some by-products can be created, and they can be used as components in the other BOM. All the issues above increase the complexity of component management and simulation modeling.

- Highly configurable customer orders. It is common for customers to choose different options and customize their automobiles. For manufacturers, each order can be different. Usually the customized cars are more expensive than standard ones. This is an opportunity, as well as a big challenge to the decision makers.
- Service level is critical. From the inbound side, the error of component supply time is not allowed to be longer than a few minutes; from the outbound side, it is mandatory for multi-national automotive manufacturers to deliver a high value product portfolio to the global market at the right time and in the right place.
- In different types of manufacturing supply chain, the decision trade-offs are different. For instance, Swaminathan et al. (1998) study three distinct domains which differ in terms of centers of decision making, heterogeneity in the supply chain, and relationship with suppliers.

We study a general four-tier automotive industry supply chain (Fig. 5.4) which consists of three suppliers, two manufacturing facilities, four distribution centers, and four customers. Figure 5.5 illustrates the snapshot of GBSE implementation of the supply chain. The main area is the logic diagram of the entire supply chain. Users can modify the diagram by dragging and dropping elements from the toolbar to the diagram. Project Navigator is in the left side of the tool, users can manage their model files and resource files in the navigator. In the left bottom, the Database Navigator is provided for checking the data in database. In the right bottom, users can modify data in the Properties Panel.

In the following sections, we describe four simulation scenarios. Each of them illustrates a typical supply chain. Twenty experiments are conducted for the four scenarios. In the model, one period is defined as one week, twelve periods are simu-

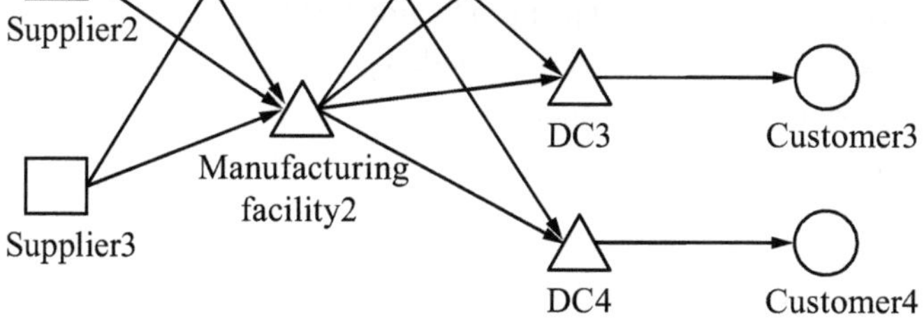

Fig. 5.4 The structure of the studied supply network

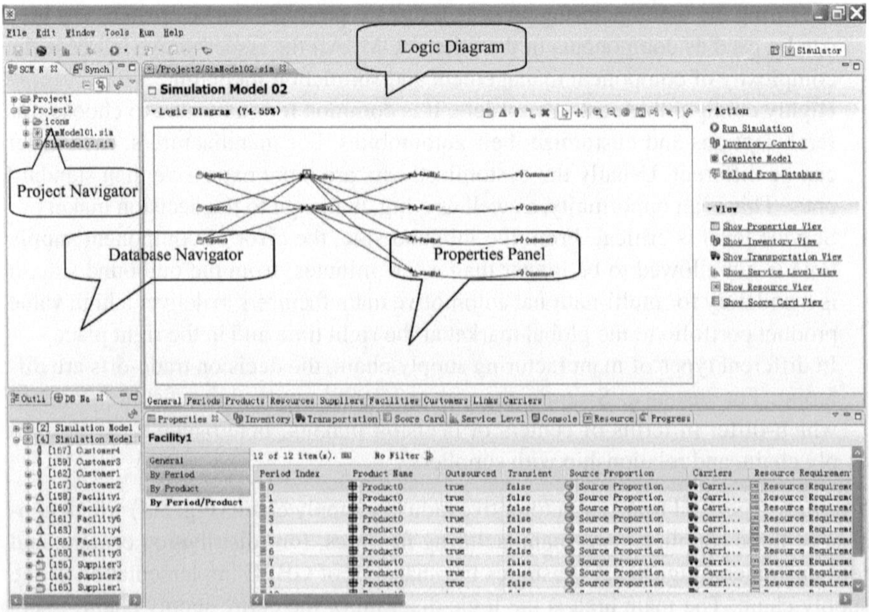

Fig. 5.5 GBSE snapshot

lated in total. A two-tier BOM is modeled. There are five finished goods with three components for each.

5.4.1 Notation

α : forecast accuracy,
σ_{d} : standard deviation of the customer demand,
μ_{d} : mean demand,
r^{o} : fill rate,
r^{m} : manufacturing failure rate,
$c^{\mathrm{b}}_{\mathrm{sp}}$: unit procurement cost for product p from supplier s,
$c^{\mathrm{m}}_{\mathrm{lp}}$: unit manufacturing cost for product p at production line l,
$c^{\mathrm{t}}_{k\mathrm{p}}$: unit transportation cost for product p with mode k,
$c^{\mathrm{v}}_{\mathrm{p}}$: unit inventory carrying cost for product p,
C^{b} : procurement cost,
C^{m} : manufacturing cost,
C^{t} : transportation cost,
C^{v} : inventory carrying cost,
$q^{\mathrm{v}}_{i\mathrm{p}}$: inventory quantity during time section i for product p,

$q^{\mathrm{b}}_{\mathrm{sp}}$: manufacturing quantity of product p from supplier s,
q^{m}_{i} : number of qualified products for manufacturing shift i,
$q^{\mathrm{m}}_{\mathrm{lp}}$: manufacturing quantity of product p at production line l,
$q^{\mathrm{t}}_{k\mathrm{p}}$: transportation quantity of product p with mode k,
q^{o}_{i} : fulfilled quantity for order i,
Q^{m}_{i} : total number of products for manufacturing shift i,
Q^{o}_{i} : ordered quantity of order i,
$t_{i\mathrm{p}}$: length of time section i of product p.

5.4.2 Scenario I: Impact of Demand Forecast Accuracy

Given different forecast accuracy levels, the fill rate of each distribution center will change accordingly. The sensitivity of GBSE can help the manager to decide if it is necessary to invest more to improve the forecast accuracy.

The forecast accuracy is defined as:

$$\alpha = 1 - \frac{\sigma_{\mathrm{d}}}{\mu_{\mathrm{d}}},$$

where α is the forecast accuracy, σ_{d} is the standard deviation of the customer demand and μ_{d} is the mean demand.

During the simulation, a set of random numbers in normal distribution is generated to represent the customer demands. According to the definition of forecast accuracy, the standard deviation of customer demand is determined as:

$$\sigma_{\mathrm{d}} = \mu_{\mathrm{d}}(1 - \alpha).$$

We run different experiments with respect to different forecast errors. Two performance metrics are collected for the study: fill rate and inventory carrying cost. The fill rate is defined as:

$$r^{\mathrm{o}} = \frac{\sum_i q^{\mathrm{o}}_i}{\sum_i Q^{\mathrm{o}}_i},$$

where r denotes the fill rate, q^{o}_i denotes the fulfilled quantity for order i and Q^{o}_i denotes the ordered quantity of order i.

In all the rest scenarios, we will use the same definition of fill rate.

For calculating the inventory carrying cost, we need to split the simulation time into a set of time sections. During each time section, the inventory keeps no change. This means each time we change the inventory, we start a new time section. The inventory carrying cost is defined as:

$$C^{\mathrm{v}} = \sum_{\mathrm{p}} \sum_{i} c^{\mathrm{v}}_{\mathrm{p}} \cdot q^{\mathrm{v}}_{i\mathrm{p}} \cdot t_{i\mathrm{p}}.$$

Where C^v denotes the inventory carrying cost, c^v_p denotes the unit inventory carrying cost for product p, q^v_{ip} denotes the inventory quantity during time period i for product p, t_{ip} denotes the length of time period i of product p.

The AS-IS forecast accuracy is 60 %. We change the forecast accuracy from 60 % to 90 %. Table 5.1 and Fig. 5.6 show the overall and individual fill rates for different values of forecast accuracy.

Table 5.1 Simulation results of scenario S1

Experiments	Forecast Accuracy(%)	Inventory Carrying Cost (K$)	Fill Rate (%)
Exp1	60	37,412	57.5
Exp2	65	31,856	65.8
Exp3	70	27,301	68.8
Exp4	75	25,609	78.8
Exp5	80	21,923	88.4
Exp6	85	21,266	89.7
Exp7	90	21,010	91.8
Exp8	95	20,726	92.0

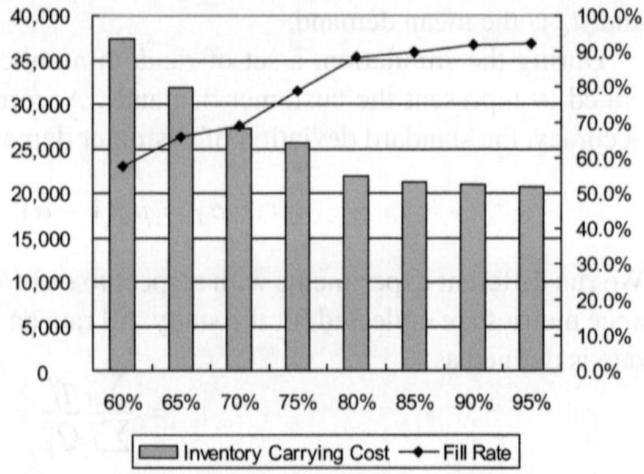

Fig. 5.6 Simulation results of scenario I

The results indicate that with an increase of forecast accuracy, the fill rate is improved significantly and, the inventory carrying cost decreases. Because less inventory is needed to ensure safety stock, lower average inventory levels are achieved. It is interesting to note that 80 % is the break point. When the forecast accuracy is lower than 80 %, the fill rate and inventory carrying cost are changed significantly. When the forecast accuracy is higher than 80 %, the magnitude of the changes diminishes. We conclude that improving the forecast accuracy is important, however, in certain extent, when the watershed is reached, forecast accuracy improvement will be more costly with considerable less return.

5.4.3 Scenario II: Impact of Supplier Selection

In automobile industry, supplier selection is very important because thousands of components are included in the BOMs. In the competitive market, each component can be supplied by different suppliers, e. g. local suppliers can supply products quickly while oversea suppliers may provide less expensive components due to low labor and raw material cost. In this example, there are three suppliers. Assume the suppliers have different procurement costs and supply lead times. Since procurement time is uncertain in nature, we will study the use of GBSE to assess how different suppliers impact the fill rate and the total procurement cost.

In this scenario, there are fifteen types of components among which ten can only be supplied by one of the three suppliers and the rest five can be supplied by all three suppliers. Supplier 3 has lowest procurement cost, however the longest lead time. Compared to Supplier 1, Supplier 2 has less lead time and relatively higher cost. We conduct four experiments for the 3 suppliers with Supplier 3 being rejected for the first two experiments and Supplier 3 being selected for the last two experiments. The proportions of procurement volume for the suppliers are different. For each experiment, fill rate and procurement cost are calculated. The procurement cost is defined as:

$$C^{\mathrm{b}} = \sum_{\mathrm{s}} \sum_{\mathrm{p}} c^{\mathrm{b}}_{\mathrm{sp}} q^{\mathrm{b}}_{\mathrm{sp}} .$$

Where C^{b} denotes the procurement cost, $c^{\mathrm{b}}_{\mathrm{sp}}$ denotes the unit procurement cost for product p from supplier s, $q^{\mathrm{b}}_{\mathrm{sp}}$ denotes the manufacturing quantity of product p from supplier s. Table 5.2 and Fig. 5.7 show the simulation results.

The result shows that if Supplier 3 is selected to provide some proportion of the supplies, the total cost decreases dramatically with the scarifying of the service level.

5.4.4 Scenario III: Impact of Different Transportation Mode

In a typical supply chain, there are usually several types of transportation modes, like air, ocean, TL (truck load), LTL (less than truck load), etc. Each transportation

Table 5.2 Simulation Results of Scenario II

Experiments	Supplier 1 %	Supplier 2 %	Supplier 3 %	Procurement Cost (K$)	Fill Rate %
Exp9	60	40	0	54,897	96.8
Exp10	40	60	0	55,104	92.7
Exp11	50	30	20	50,741	86.0
Exp12	30	50	20	51,211	81.3

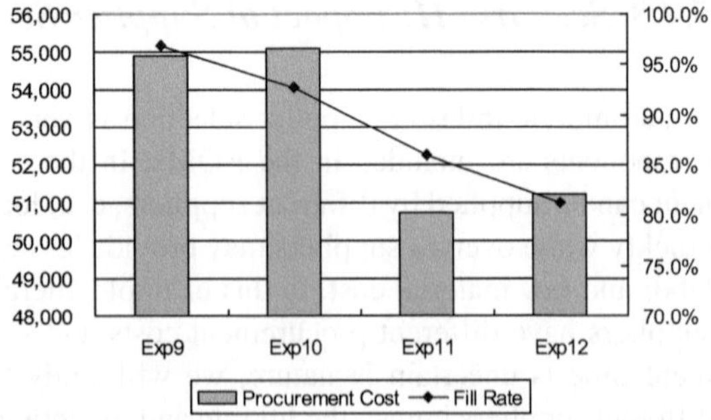

Fig. 5.7 Simulation results of scenario II

mode has different elements of uncertainty. The decision makers need to answer questions like what transportation modes should be selected, how to assign usage ratio for each transportation mode, etc. There are three types of transportation in each supply chain: inbound transportation time, internal transportation time, and outbound transportation time. Inbound transportation is from suppliers to internal facilities; outbound transportation is from internal facilities to customers or from customer to other customers; internal transportation is between internal facilities.

In this example, air and ocean modes are supported for inbound transportation. Since the air transportation is expensive, it is used for high priority or urgent orders only. The cost of ocean transportation is much lower, and it is proper for normal shipping. But the lead time of ocean mode is usually several weeks, so it's not flexible and a precise procurement plan needs to be made in advance. Truck and train modes are supported for internal and outbound transportation. Both modes need to follow transportation schedules, but the truck mode is more flexible. Table 5.3 summarizes the characteristics of the transportation modes.

Table 5.3 Comparison of different transportation modes

Transportation Mode	Cost	Transportation Time
Air	Very high	A few days
Ocean	Very low	Several weeks
Truck	High	Several days
Train	Low	Several days

We conduct four experiments for this scenario and calculate the transportation cost and fill rate. The transportation cost is defined as:

$$C^{\mathrm{t}} = \sum_{k} \sum_{\mathrm{p}} c^{\mathrm{t}}_{k\mathrm{p}} q^{\mathrm{t}}_{k\mathrm{p}} \,.$$

Where C^{t} denotes the transportation cost, $c_{k\mathrm{p}}^{\mathrm{t}}$ denotes the unit transportation cost for product p with mode k, $q_{k\mathrm{p}}^{\mathrm{t}}$ denotes the transportation quantity of product p with mode k. Table 5.4 and Fig. 5.8 show the simulation results.

Table 5.4 Simulation results of scenario III

Experiments	Air %	Ocean %	Truck %	Train %	Transportation Cost (K$)	Fill Rate %
Exp13	10	90	30	70	38,322	82.3
Exp14	10	90	70	30	39,210	85.5
Exp15	20	80	30	70	43,687	91.1
Exp16	20	80	70	30	44,471	96.3

Fig. 5.8 Simulation Results of Scenario III

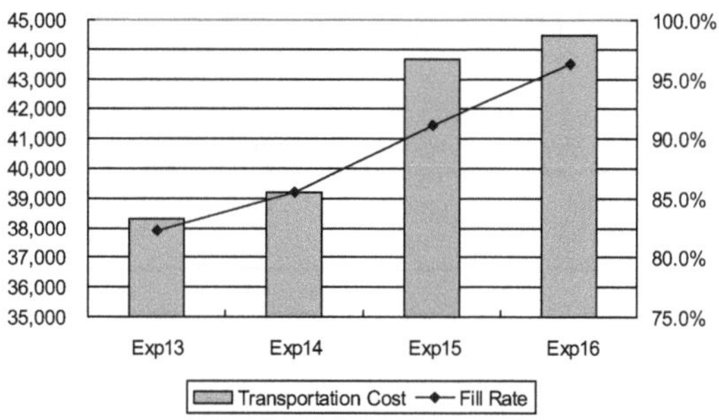

From the results, we conclude that the use of air transportation instead of ocean for the inbound will encounter higher cost, and same for the use of truck instead of train, however, the fill rate will be improved significantly.

5.4.5 Scenario IV: Impact of Quality Uncertainty

There are usually multiple production lines in an automotive manufacturing facility. Two key attributes of production lines are failure rate and unit production cost. The failure rate is related to product quality, it is defined as:

$$r^{\mathrm{m}} = 1 - \frac{\sum_i q_i^{\mathrm{m}}}{\sum_i Q_i^{\mathrm{m}}}.$$

Where r^{m} denotes the failure rate, q_i^{m} denotes the number of qualified products for manufacturing shift i, Q_i^{m} denotes the total number of products for manufacturing shift i.

In this experiment, we focus on manufacturing facility1 which consists of three existing production lines (PL1, PL2, and PL3). PL1 and PL2 are the same. PL3 has lower failure rate, while the operation cost is relatively higher than PL1 and PL2. Meanwhile, the management team is considering if it's required to purchase two new production lines (PL4 and PL5). They have lowest failure rate and highest operation cost.

Four experiments are designed for this study in which PL4 and PL5 are not purchased with only PL1, PL2 and PL3 are used for the first two experiments, all give production lines are used for the last two experiments. Manufacturing cost and fill rate are calculated for each experiment

$$C^{\mathrm{m}} = \sum_{\mathrm{l}} \sum_{\mathrm{p}} c_{\mathrm{lp}}^{m} q_{\mathrm{lp}}^{\mathrm{m}}.$$

Where C^{m} denotes the manufacturing cost, $c_{\mathrm{lp}}^{\mathrm{m}}$ denotes the unit manufacturing cost for product p at production line l, $q_{\mathrm{lp}}^{\mathrm{m}}$ denotes the manufacturing quantity of product p at production line l. Table 5.5 and Fig. 5.9 summarize the simulation results.

Table 5.5 Simulation results of scenario IV

Experiments	PL1 %	PL2 %	PL3 %	PL4 %	PL5 %	Manufacturing Cost (K$)	Fill Rate %
Exp17	40	40	20	0	0	46,732	82.3
Exp18	30	30	40	0	0	47,561	86.0
Exp19	20	20	20	20	20	48,785	93.1
Exp20	10	10	20	30	30	49,760	95.9

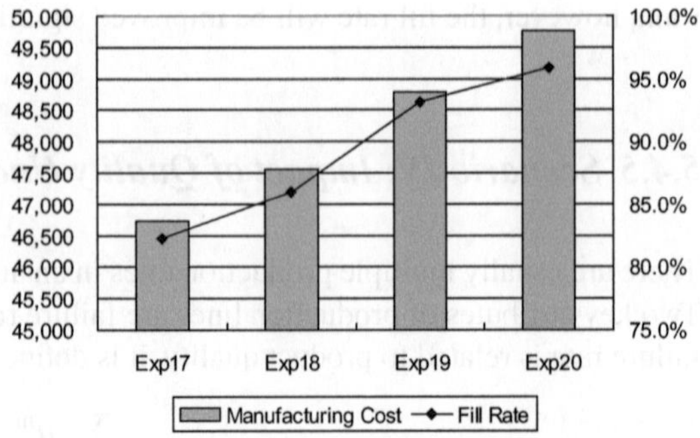

Fig. 5.9 Simulation results of scenario IV

As shown by the results, after introducing two new production lines, although the total cost increases, the fill rates are improved significantly. In addition to this, if more products are produced with PL3, the fill rate will also get improved.

5.5 Conclusion

In this chapter, we introduce GBSE, an integrated supply chain simulation environment developed by IBM China Research Lab. It has been applied for both IBM internal and external supply chains to evaluate several key performance metrics and optimize supply network operations. The tool is designed for tactical-level of decision making, and it supports wide range of supply chain processes. Simulation analysts can use it to conduct several kinds of what-if analysis and risk analysis in supply chain context, so as to make correct decisions more effectively. In this study, we demonstrate the applicability of GBSE to responsive supply chain management by presenting four scenarios to study varied uncertainties in forecasting, various decisions in selecting suppliers, transportation modes and production lines. We conclude that GBSE can facilitate the decision making by analyzing the impact of different uncertainties in details quantitatively. While promising, we will design experiments with more impacting factors being considered and further in-depth analysis will be conducted.

References

1. B.P. Gan, L. Liu, S. Jain, S.J. Turner, W. Cai, and W.J. Hsu. "Distributed supply chain simulation across enterprise boundaries". Proceedings of the 2000 Winter Simulation Conference, ed. by J.A. Joines, R.R. Barton, K. Kang, and P.A. Fishwick, 1245–1251. 2000
2. C.F. Sabel, , H. Kern, and G. Herrigel. Collaborative manufacturing: New supplier relations in the automobile industry and the redefinition of the industrial corporation. IMVP International Policy Forum, MIT, Cambridge, MA. 1989
3. J. Dong, H. Ding, C. Ren, and W. Wang. IBM SmartSCOR – A SCOR Based Supply Chain Transfor-mation Platform, Proceedings of Winter Simulation Conference, 2006
4. G. Archibald, N. Karabakal, and P. Karlsson. "Supply chain vs. supply chain: Using simulation to compete beyond the four walls", Proceedings of the 1999 Winter Simulation Conference, ed. by Farrington, H.B. Nembhard, D.T. Sturrock, and G.W. Evans, 888–896. 1999
5. J. Banks, J.S. Carson, B.L. Nelson, and D.M. Nicol. Discrete-event Simulation, Prentice-Hall, Upper Saddle River, NJ, New Jersey: Prentice-Hall, Inc. 2005
6. J.M. Swaminathan, S.F. Smith, and N.M. Sadeh. Modeling Supply Chain Dynamics: A Multiagent Approach, Decision Sciences. 1998
7. J. Venkateswaran, Y.-J. Son, and B. Kulvatunyou. "Investigation of influence of modeling fidelities on supply chain dynamics". Proceedings of the 2002 Winter Simulation Conference, ed. by E. Yücesan, C.-H. Chen, J.L. Snowdon, and J.M. Charnes, 1183–1191. 2002
8. J. Blackhurst, T. Wu, and P. O'Grady, "Network-based approach to modeling uncertainty in a supply chain", International Journal of Production Research, 42(8):1639–1658. 2004
9. J.P.C. Kleijnen, Supply Chain Simulation: A Survey, International Journal of Simulation and Process Modeling, 2003
10. L. Chwif, M.R.P. Barretto, and E. Saliby. "Supply chain analysis: Spreadsheet or simulation?" Proceedings of the 2002 Winter Simulation Conference, ed. by E. Yücesan, C.-H. Chen, J.L. Snowdon, and J.M. Charnes, 59–66. 2002
11. J. Liu, W. Wang, Y. Chai, Y. Liu. "Easy-SC: A Supply Chain Simulation Tool", Proceedings of Winter Simulation Conference, 2004

12. M. Hennessee. "Challenges facing global supply chains in the 21st century", Proceedings of the 1998 Winter Simulation Conference, ed. by D.J. Medeiros, E.F. Watson, J.S. Carson, and M.S. Manivannan, 3–4. 1998
13. P. Lendermann, N. Julka, B.P. Gan, D. Chen, L.F. McGinnis, and J.P. McGinnis. "Distributed supply chain simulation as a decision-support tool for the semiconductor industry", Simulation, 79(3):126–138. 2003
14. Repast, available via http://repast.sourceforge.net/, 2008
15. M.D. Rossetti and H.T. Chan. "A Prototype Object-Oriented Supply Chain Simulation Framework", Proceedings of Winter Simulation Conference, 2003
16. M.D. Rossetti, M. Miman, V. Varghese, and Y. Xiang. "An Object-Oriented Framework for Simulating Multi-echelon Inventory Systems", Proceedings of Winter Simulation Conference, 2006
17. S. Bagchi, S.J. Buckley, M. Ettl, and G.Y. Lin. "Experience using the IBM supply chain simulator", Proceedings of the 1998 Winter Simulation Conference, ed. by D.J. Medeiros, E.F. Watson, J.S. Carson, and M.S. Manivannan, 1387–1394. 1998
18. S. Jain, C.C. Lim, B.P. Gan, and Y.H. Low. "Criticality of detailed modeling in semiconductor supply chain simulation", Proceedings of the 1999 Winter Simulation Conference, ed. by P.A. Farrington, H.B. Nembhard, D.T. Sturrock, and G.W. Evans, 888–896. 1999
19. S. Jain, E.C. Ervin, A.P. Lathrop, R.W. Workman, L.M. Collins. "Analyzing the supply chain for a large logistics operation using simulation". Proceedings of the 2001 Winter Simulation Conference, ed. by B.A. Peters, J.S. Smith, D.J. Medeiros, and M.W. Rohrer, 1123–1128. 2001
20. S. Jain, N.F. Choong, and W. Lee. "Modeling computer assembly operations for supply chain integration", Proceedings of the 2002 Winter Simulation Conference, ed. by E. Yücesan, C.-H. Chen, J.L. Snowdon, and J.M. Charnes, 1165–1173. 2002
21. S.T. Enns, and P. Suwanruji. "A simulation test bed for production and supply chain modeling", Proceedings of the 2003 Winter Simulation Conference, ed. by S. Chick, P.J. Sánchez, D. Ferrin, and D.J. Morrice, 1174–1182. IEEE. 2003
22. Scalable Simulation Framework (SSF), available via http://www.ssfnet.org/homePage.html, 2008
23. Supply Chain Guru, available via http://www.llamasoft.com/, 2008
24. T. Wu, P. O'Grady, "A Methodology for Improving the Design of a Supply Chain", *International Journal of Computer Integrated Manufacturing*, 17(4):281–293, 2004

Section II
Decision Making and Risk Mitigation in the Supply Chain

Chapter 6
Modeling of Supply Chain Risk Under Disruptions with Performance Measurement and Robustness Analysis

Qiang Qiang, Anna Nagurney, and June Dong

Abstract In this chapter, we develop a new supply chain network model with multiple decision-makers associated at different tiers and with multiple transportation modes for shipment of the good between tiers. The model formulation captures supply-side risk as well as demand-side risk, along with uncertainty in transportation and other costs. The model also incorporates the individual attitudes towards disruption risks among the manufacturers and the retailers, with the demands for the product associated with the retailers being random. We present the behavior of the various decision-makers, derive the governing equilibrium conditions, and establish the finite-dimensional variational inequality formulation. We also propose a weighted supply chain performance and robustness measure based on our recently derived network performance/efficiency measure and provide supply chain examples for which the equilibrium solutions are determined along with the robustness analyses. This chapter extends previous supply chain research by capturing supply-side disruption risks, transportation and other cost risks, and demand-side uncertainty within an integrated modeling and robustness analysis framework.

6.1 Introduction

Supply chain disruptions and the associated risk are major topics in theoretical and applied research, as well as in practice, since risk in the context of supply chains may be associated with the production/procurement processes, the transportation/shipment of the goods, and/or the demand markets. In fact, Craighead, et al. (2007) have argued that supply chain disruptions and the associated operational and financial risks are the most pressing issue faced by firms in today's competitive global environment. Notably, the focus of research has been on "demand-side" risk, which is related to fluctuations in the demand for products, as opposed to the "supply-side" risk, which deals with uncertain conditions that affect the production and transportation processes of the supply chain. For a discussion of the distinction between these two types of risk, see Snyder (2003).

T. Wu, J. Blackhurst (eds.), *Managing Supply Chain Risk and Vulnerability*

For example, several recent major disruptions and the associated impacts on the business world have vividly demonstrated the need to address supply-side risk with a case in point being a fire in the Phillips Semiconductor plant in Albuquerque, New Mexico, causing its major customer, Ericsson, to lose $400 million in potential revenues. On the other hand, another major customer, Nokia, managed to arrange alternative supplies and, therefore, mitigated the impact of the disruption (cf. Latour, 2001). Another illustrative example concerns the impact of Hurricane Katrina, with the consequence that 10 %–15 % of total U.S. gasoline production was halted, which not only raised the oil price in the U.S., but also overseas (see, e.g., Canadian Competition Bureau, 2006). Moreover, the world price of coffee rose 22 % after Hurricane Mitch struck the Central American republics of Nicaragua, Guatemala, and Honduras, which also affected supply chains worldwide (Fairtrade Foundation, 2002). As summarized by Sheffi (2005) on page 74, one of the main characteristics of disruptions in supply networks is "the seemingly unrelated consequences and vulnerabilities stemming from global connectivity." Indeed, supply chain disruptions may have impacts that propagate not only locally but globally and, hence, a holistic, system-wide approach to supply chain network modeling and analysis is essential in order to be able to capture the complex interactions among decision-makers.

Indeed, rigorous modeling and analysis of supply chain networks, in the presence of possible disruptions is imperative since disruptions may have lasting major financial consequences. Hendricks and Singhal (2005) analyzed 800 instances of supply chain disruptions experienced by firms whose stocks are publicly traded. They found that the companies that suffered supply chain disruptions experienced share price returns 33 % to 40 % lower than the industry and the general market benchmarks. Furthermore, share price volatility was 13.5 % higher in these companies in the year following a disruption than in the prior year. Based on their findings, it is evident that only well-prepared companies can effectively cope with supply chain disruptions. Wagner and Bode (2007), in turn, designed a survey to empirically study the responses from executives of firms in Germany regarding their opinions as to the factors that impact supply chain vulnerability. The authors found that demand-side risks are related to customer dependence while supply-side risks are associated with supplier dependence, single sourcing, and global sourcing.

The goal of supply chain risk management is to alleviate the consequences of disruptions and risks or, simply put, to increase the *robustness* of a supply chain. However, there are very few quantitative models for measuring supply chain robustness. For example, Bundschuh, et al. (2003) discussed the design of a supply chain from both reliability and robustness perspectives. The authors built a mixed integer programming supply chain model with constraints for reliability and robustness. The robustness constraint was formulated in an implicit form: by requiring the suppliers' sourcing limit to exceed a certain level. In this way, the model built redundancy into a supply chain. Snyder and Daskin (2005) examined supply chain disruptions in the context of facility location. The objective of their model was to select locations for warehouses and other facilities that minimize the transportation costs to customers and, at the same time, account for possible closures of facilities that would result in re-routing of the product. However, as commented in Snyder and Shen (2006), "Although these are multi-location models, they focus primarily on the local effects of disruptions." Santoso, et al., 2005 applied a sample average approximation scheme to study the stochastic facility location problem by considering different disruption scenarios.

Tang (2006a) also discussed how to deploy certain strategies in order to enhance the robustness and the resiliency of supply chains. Kleindorfer and Saad (2005), in turn, provided an overview of strategies for mitigating supply chain disruption risks, which were exemplified by a case study in a chemical product supply chain. For a comprehensive review of supply chain risk management models to that date, please refer to Tang (2006b).

To-date, however, most supply disruption studies have focused on a local point of view, in the form of a single-supplier problem (see, e. g., Gupta, 1996; Parlar, 1997) or a two-supplier problem (see, e. g., Parlar and Perry, 1996). Very few papers have examined supply chain risk management in an environment with multiple decision-makers and in the case of uncertain demands (cf. Tomlin, 2006). We believe that it is imperative to study supply chain risk management from a holistic point of view and to capture the interactions among the multiple decision-makers in the various supply chain network tiers. Indeed, such a perspective has also been argued by Wu et al. (2006), who focused on inbound supply risk analysis. Towards that end, in this chapter, we take an entirely different perspective, and we consider, for the first time, supply chain robustness in the context of multi-tiered supply chain networks with multiple decision-makers under equilibrium conditions. For a plethora of supply chain network equilibrium models and the associated underlying dynamics, see the book by Nagurney (2006a).

Of course, in order to study supply chain robustness, an informative and effective performance measure is first required. Beamon (1998, 1999) reviewed the supply chain literature and suggested directions for research on supply chain performance measures, which should include criteria on efficient resource allocation, output maximization, and flexible adaptation to the environmental changes (see also, Lee and Whang, 1999; Lambert and Pohlen, 2001; Lai et al., 2002). We emphasize that different supply performance measures can be devised based on the specific nature of the problem. In any event, the discussion here is not meant to cover all the existing supply chain performance measures. Indeed, we are well aware that it is a daunting task to propose a supply chain performance measure that covers all aspects of supply chains. We believe that such a discussion will be an ongoing research topic for decades to follow. In this chapter, we study supply chain robustness based on a novel network performance measure proposed by Qiang and Nagurney (2008), which captures the network flows, the costs, and the decision-makers' behavior under network equilibrium conditions.

In particular, the model developed in this chapter extends the supply chain model of Nagurney et al. (2002) with consideration of random demand (cf. Nagurney et al., 2005). In order to study supply chain robustness, the new model contains the following novel features:

- We associate each process in a supply chain with random cost parameters to represent the impact of disruptions to the supply chain.
- We extend the aforementioned supply chain models to capture the attitude of the manufacturers and the retailers towards disruption risks.
- We propose a weighted performance measure to evaluate different supply chain disruptions.
- Different transportation modes are considered in the model (see also, e. g., Dong et al., 2002; Dong et al., 2005). In the multimodal transportation supply chain, alternative transportation modes can be used in the case of the failure of a transportation mode. Indeed, many authors have emphasized that redundancy needs to be considered in the design of supply chains in order to prevent supply chain disruptions. For example, Wilson (2007) used a system dynamic simulation to study the relationship between transportation disruptions and supply chain performance. The author found that the existence of transportation alternatives significantly improved supply chain performance in the case of transportation disruptions.

In this chapter, we assume that the probability distributions of the disruption related cost parameters are known. This assumption is not unreasonable given today's advanced information technology and increasing awareness of the risks among managers. A great deal of disruption related information can be obtained from a careful examination and abstraction of the relevant data sources. Specifically, as indicated by Sheffi (2005, p. 55), "… as investigation boards and legal proceedings have revealed, in many cases relevant data are on the record but not funneled into a useful place or not analyzed to bring out the information in the data". Moreover, Holmgren (2007) also discussed ways to improve prediction of disruptions, using, for example: historical data analysis, mathematical modeling, and expert judgments. Furthermore, we assume that the random cost parameters are independent.

The organization of this chapter is as follows. In Sect. 6.2, we present the model of a supply chain network faced with (possible) disruptions and in the case of random demands and multiple transportation modes. In Sect. 6.3, we provide a definition of a weighted supply chain performance measure with consideration of robustness. In Sect. 6.4, we present numerical examples in order to illustrate the model and concepts introduced in this chapter. The chapter concludes with Sect. 6.5, which summarizes the results obtained and provides suggestions for future research.

6.2 The Supply Chain Model with Disruption Risks and Random Demands

The topology of the supply chain network is depicted in Fig. 6.1.

The supply chain model consists of m manufacturers, with a typical manufacturer denoted by i, n retailers with a typical retailer denoted by j, and o demand markets with a typical demand market denoted by k. Furthermore, we assume that there are

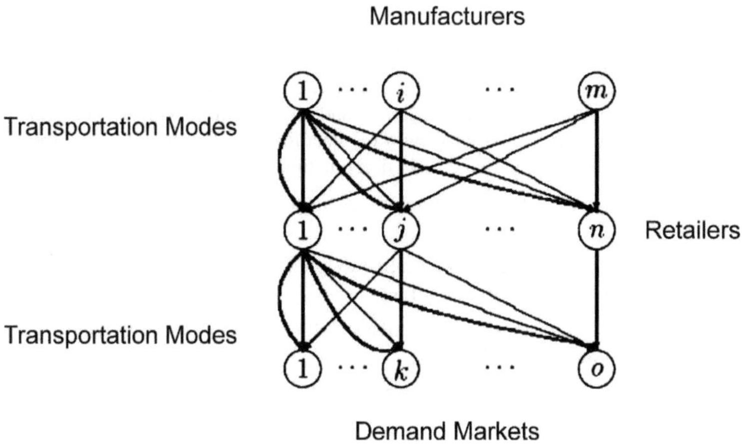

Fig. 6.1 The multitiered network structure of the supply chain

g transportation modes from manufacturers to retailers, with a typical mode denoted by u and there are h transportation modes between retailers and demand markets, with a typical mode denoted by v. Typical transportation modes may include trucking, rail, air, sea, etc. By allowing multiple modes of transportation between successive tiers of the supply chain we also generalize the earlier models of Dong et al. (2002) and Dong et al. (2005).

Manufacturers are assumed to produce a homogeneous product, which can be purchased by retailers, who, in turn, make the product available to demand markets. Each process in the supply chain is associated with some random parameters that affect the cost functions. The relevant notation is summarized in Table 6.1.

Table 6.1 Notation for the supply chain network model

Notation	Definition
q	m-dimensional vector of the manufacturers' production outputs with components: $q_1, \ldots, q_m$
Q^1	mng-dimensional vector of product shipments between manufacturers and retailers via the transportation modes with component iju denoted by q_{ij}^u
Q^2	noh-dimensional vector of product shipments between retailers and demand markets via the transportation modes with component jkv denoted by q_{jk}^v
α	m-dimensional vector of nonnegative random parameters with being the random parameter associated with the product cost of manufacturer i and the corresponding cumulative distribution function is given by $F_i(\alpha_i)$
β	mng-dimensional vector of nonnegative random parameters with β_{ij}^u being the random parameter associated with the transportation cost of manufacturer i and retailer j via mode u and the corresponding cumulative distribution function is given by $F_{ij}^u(\beta_{ij}^u)$

Table 6.1 (continued)

Notation	Definition
η	n-dimensional vector of nonnegative random parameters with η_j being the random parameter associated with the handling cost of retailer j and the corresponding cumulative distribution function is given by $F_j(\eta_j)$
γ	n-dimensional vector of shadow prices associated with the retailers with component j denoted by γ_j
θ	m-dimensional vector of nonnegative weights with θ_i reflecting manufacturer i's attitude towards disruption risks
ϖ	n-dimensional vector of nonnegative weights with ϖ_j reflecting retailer j's attitude towards disruption risks
$f_i(q,\alpha_i) \equiv f_i(Q^1,\alpha_i)$	Production cost of manufacturer i with random parameter
$\hat{F}(q) \equiv \hat{F}(Q^1)$	Expected production cost function of manufacturer i with marginal production cost with respect to q_{ij}^u denoted by $\frac{\partial \hat{F}_i(Q^1)}{\partial q_{ij}^u}$
$VF_i(Q^1)$	Variance of the production cost of manufacturer i with marginal with respect to q_{ij}^u denoted by $\frac{\partial VF_i(Q^1)}{\partial q_{ij}^u}$
$c_{ij}^u\left(q_{ij}^u,\beta_{ij}^u\right)$	Transaction cost between manufacturer i and retailer j via transportation mode u with the random parameter β_{ij}^u
$\hat{c}_{ij}^u\left(q_{ij}^u\right)$	Expected transaction cost between manufacturer i and retailer j via transportation mode u with marginal transaction cost denoted by $\frac{\partial \hat{c}_{ij}^u\left(q_{ij}^u\right)}{\partial q_{ij}^u}$
$VC_{ij}^u\left(q_{ij}^u\right)$	Variance of the transaction cost between manufacturer i and retailer j via transportation mode u with marginal denoted by $\frac{\partial VC_{ij}^u\left(q_{ij}^u\right)}{\partial q_{ij}^u}$
$C_j(Q^1,Q^2,\eta_j)$	Handling cost of retailer j with random parameter η_j
$\hat{C}_j^1(Q^1,Q^2)$	Expected handling cost of retailer j with marginal handling cost with respect to q_{ij}^u denoted by $\frac{\partial \hat{C}_j^1(Q^1,Q^2)}{\partial q_{ij}^u}$ and the marginal handling cost with respect to q_{jk}^v denoted by $\frac{\partial \hat{C}_j^1(Q^1,Q^2)}{\partial q_{jk}^v}$
$VC_j^1(Q^1,Q^2)$	Variance of the handling cost of retailer j with marginal with respect to q_{ij}^u denoted by $\frac{\partial VC_j^1(Q^1,Q^2)}{\partial q_{ij}^u}$ and the marginal with respect to q_{jk}^v denoted by $\frac{\partial VC_j^1(Q^1,Q^2)}{\partial q_{jk}^v}$
$c_{jk}^v(Q^2)$	Unit transaction cost between retailer j and demand market k via transportation mode v
$d_k(\rho_3)$	Random demand at demand market k with expected value $\hat{d}_k(\rho_3)$
ρ_3	Vector of prices of the product at the demand markets with ρ_{3k} denoting the demand price at demand market k

6.2.1 The Behavior of the Manufacturers

We assume a homogeneous product economy meaning that all manufacturers produce the same product which is then shipped to the retailers, who, in turn, sell the product to the demand markets.

Since the total amount of the product shipped from a manufacturer via different transportation modes has to be equal to the amount of the production of each manufacturer, we have the following relationship between the production of manufacturer i and the shipments to the retailers:

$$q_i = \sum_{j=1}^{n} \sum_{u=1}^{g} q_{ij}^u, \quad i = 1, \ldots, m. \tag{6.1}$$

We assume that disruptions will affect the production processes of manufacturers, the impact of which is reflected in the production cost functions. For each manufacturer i, there is a random parameter α_i that reflects the impact of disruption to his production cost function. The expected production cost function is given by:

$$\hat{F}_i(Q^1) \equiv \int_{\alpha_i} f_i\left(Q^1, \alpha_i\right) \mathrm{d}F_i(\alpha_i), \quad i = 1, \ldots, m. \tag{6.2}$$

We further denote the variance of the above production cost function as $VF_i(Q^1)$ where $i = 1, \ldots, m$.

As noted earlier, we assume that each manufacturer has g types of transportation modes available to ship the product to the retailers, the cost of which is also subject to disruption impacts. The expected transportation cost function is given by:

$$\hat{C}_{ij}^u\left(q_{ij}^u\right) \equiv \int_{\beta_{ij}^u} c_{ij}^u\left(q_{ij}^u, \beta_{ij}^u\right) \mathrm{d}F_{ij}^u\left(\beta_{ij}^u\right),$$

$$i = 1, \ldots, m; \quad j = 1, \ldots, n; \quad u = 1, \ldots, g. \tag{6.3}$$

We further denote the variance of the above transportation cost function as $VC_{ij}^u(Q^1)$ where $i = 1, \ldots, m$; $j = 1, \ldots, n$; $u = 1, \ldots, g$.

It is well-known in economics that variance may be used to measure risk (see, e.g., Silberberg and Suen, 2000; Tomlin, 2006 using such an approach to study risks in applications to supply chains). Therefore, we assign a nonnegative weight θ_i to the variance of the cost functions for each manufacturer to reflect his attitude towards disruption risks. The larger the weight is, the larger the penalty a manufacturer imposes on the risk, and, therefore, the more risk-averse the manufacturer is.

Let ρ_{1ij}^{u*} denote the price charged for the product by manufacturer i to retailer j when the product is shipped via transportation mode u. Hence, manufacturers can price according to their locations as well as according to the transportation modes utilized. Each manufacturer faces two objectives: to maximize his expected profit

and to minimize the disruption risks adjusted by his risk attitude. Therefore, the objective function for manufacturer i; $i = 1, \ldots, m$ can be expressed as follows:
Maximize

$$\sum_{j=1}^{n}\sum_{u=1}^{g} \rho_{1ij}^{u*} q_{ij}^{u} - \hat{F}_i(Q^1) - \sum_{j=1}^{n}\sum_{u=1}^{g} \hat{C}_{ij}^{u}\left(q_{ij}^{u}\right)$$
$$-\theta_i \left[VF_i(Q^1) + \sum_{j=1}^{n}\sum_{u=1}^{g} VC_{ij}^{u}\left(q_{ij}^{u}\right) \right] \quad (6.4)$$

subject to:

$$q_{ij}^{u} \geq 0, \quad \text{for all } i,\ j, \text{ and } u\,.$$

The first term in (6.4) represents the revenue. The second term is the expected disruption related production cost. The third term is the expected disruption related transportation cost. The fourth term is the cost of disruption risks adjusted by each manufacturer's attitude.

We assume that, for each manufacturer, the production cost function and the transaction cost function without disruptions are continuously differentiable and convex. It is easy to verify that $\hat{F}_i(Q^1)$, $VF_i(Q^1)$, $\hat{C}_{ij}^{u}\left(q_{ij}^{u}\right)$, and $VC_{ij}^{u}\left(q_{ij}^{u}\right)$ are also continuously differentiable and convex. Furthermore, we assume that manufacturers compete in a non-cooperative fashion in the sense of Nash (1950, 1951). Hence, the optimality conditions for all manufacturers simultaneously (cf. Bazaraa et al., 1993; Nagurney, 1999) can be expressed as the following variational inequality: determine $Q^{1*} \in R_{+}^{mng}$ satisfying:

$$\sum_{i=1}^{m}\sum_{j=1}^{n}\sum_{u=1}^{g} \left[\frac{\partial \hat{F}_i\left(Q^{1*}\right)}{\partial q_{ij}^{u}} + \frac{\partial \hat{C}_{ij}^{u}\left(q_{ij}^{u*}\right)}{\partial q_{ij}^{u}} \right.$$
$$\left. +\theta_i \left(\frac{\partial VF_i\left(Q^{1*}\right)}{\partial q_{ij}^{u*}} + \frac{\partial VC_{ij}^{u}\left(q_{ij}^{u*}\right)}{\partial q_{ij}^{u*}} \right) - \rho_{1ij}^{u*} \right]$$
$$\times \left[q_{ij}^{u} - q_{ij}^{u*} \right] \geq 0, \quad \forall Q^1 \in R_{+}^{mng}\,. \quad (6.5)$$

6.2.2 The Behavior of the Retailers

The retailers, in turn, are involved in transactions both with the manufacturers and the demand markets since they must obtain the product to deliver to the consumers at the demand markets.

Let ρ_{2jk}^{v*} denote the price charged for the product by retailer j to demand market k when the product is shipped via transportation mode v. Hence, retailers can price according to their locations as well as according to the transportation modes utilized. This price is determined endogenously in the model along with the prices associated with the manufacturers, that is, the ρ_{1ij}^{u*}, for all i, j and u. We assume that certain disruptions will affect the retailers' handling processes (e. g., the storage and display processes). An additional random risk/disruption related random parameter η_j is associated with the handling cost of retailer j. Recall that we also assume that there are h types of transportation modes available to each retailer for shipping the product to the demand markets. The expected handling cost is given by:

$$\hat{C}_j^1\left(Q^1, Q^2\right) \equiv \int_{\eta_j} c_j\left(Q^1, Q^2, \eta_j\right) \mathrm{d}F_j\left(\eta_j\right), \quad j = 1, \ldots, n. \tag{6.6}$$

We further denote the variance of the above handling cost function as $VC_j^1\left(Q^1, Q^2\right)$ where $j = 1, \ldots, n$.

Furthermore, similar to the case for the manufacturers, we associate a nonnegative weight ϖ_j to the variance of each retailer's handling cost according to his attitude towards risk. Each retailer faces two objectives: to maximize his expected profit and to minimize the disruption risks adjusted by his risk attitude. Therefore, the objective function for

retailer j; $j = 1, \ldots, n$ can be expressed as follows:
Maximize

$$\sum_{k=1}^{o} \sum_{v=1}^{h} \rho_{2jk}^{v*} q_{jk}^{v} - \hat{C}_j^1\left(Q^1, Q^2\right) - \sum_{i=1}^{m} \sum_{u=1}^{g} \rho_{1ij}^{u*} q_{ij}^{u} - \varpi_j VC_j^1\left(Q^1, Q^2\right) \tag{6.7}$$

subject to:

$$\sum_{k=1}^{o} \sum_{v=1}^{h} q_{jk}^{v} \leq \sum_{i=1}^{m} \sum_{u=1}^{g} q_{ij}^{u} \tag{6.8}$$

and the nonnegativity constraints: $q_{ij}^{u} \geq 0$ for all i, j, and u; $q_{jk}^{v} \geq 0$ for all j, k, and v.

Objective function (6.7) expresses that the difference between the revenues minus the expected handling cost, the payout to the manufacturers and the weighted disruption risk is to be maximized. Constraint (6.8) states that retailers cannot purchase more product from a retailer than is available in stock.

As noted in Table 6.1, γ_j is the Lagrange multiplier associated with constraint (6.6) for retailer j. Furthermore, we assume that, for each retailer, the handling cost without disruptions is continuously differentiable and convex. It is easy to verify that $\hat{C}_j^1\left(Q^1, Q^2\right)$ and $VC_j^1\left(Q^1, Q^2\right)$ are also continuously differentiable and convex. We assume that retailers compete with one another in a noncooperative manner, seeking to determine their optimal shipments from the manufacturers and to the demand markets. The optimality conditions for all retailers simultaneously coincide with the solution of the following variational inequality: determine $\left(Q^{1*}, Q^{2*}, \gamma^*\right) \in R_{+}^{mng+noh+n}$ satisfying:

$$\sum_{i=1}^{m}\sum_{j=1}^{n}\sum_{u=1}^{g}\left[\frac{\partial \hat{C}_j^1\left(Q^{1*},Q^{2*}\right)}{\partial q_{ij}^u}+\rho_{1ij}^{u*}+\varpi_j\frac{\partial VC_j^1\left(Q^{1*},Q^{2*}\right)}{\partial q_{ij}^u}-\gamma_j^*\right]$$
$$\times\left[q_{ij}^u-q_{ij}^{u*}\right]$$
$$+\sum_{j=1}^{n}\sum_{k=1}^{o}\sum_{v=1}^{h}\left[-\rho_{2jk}^{v*}+\gamma_j^*+\frac{\partial \hat{C}_j^1\left(Q^{1},Q^{2}\right)}{\partial q_{jk}^v}+\varpi_j\frac{\partial VC_j^1\left(Q^{1*},Q^{2*}\right)}{\partial q_{jk}^v}\right]$$
$$\times\left[q_{jk}^v-q_{jk}^{v*}\right]$$
$$+\sum_{j=1}^{n}\left[\sum_{i=1}^{m}\sum_{u=1}^{g}q_{ij}^{u*}-\sum_{k=1}^{o}\sum_{v=1}^{h}q_{jk}^{v*}\right]\times\left[\gamma_j-\gamma_j^*\right]$$
$$\geq 0,\quad \forall\left(Q^1,Q^2,\gamma\right)\in R_+^{mng+noh+n}. \tag{6.9}$$

6.2.3 The Market Equilibrium Conditions

We now turn to a discussion of the market equilibrium conditions. Subsequently, we construct the equilibrium condition for the entire supply chain network.

The equilibrium conditions associated with the product shipments that take place between the retailers and the consumers are the *stochastic economic equilibrium* conditions, which, mathematically, take on the following form: for any retailer with associated demand market k; $k = 1, \ldots, o$:

$$\hat{d}_k\left(\rho_3^*\right)\begin{cases}\leq \sum_{j=1}^{o}\sum_{v=1}^{h} q_{jk}^{v*}, & \text{if } \rho_{3k}^*=0,\\ = \sum_{j=1}^{o}\sum_{v=1}^{h} q_{jk}^{v*}, & \text{if } \rho_{3k}^*>0,\end{cases} \tag{6.10a}$$

$$\rho_{2jk}^{v*}+c_{jk}^v\left(Q^{2*}\right)\begin{cases}\geq \rho_{3k}^*, & \text{if } q_{jk}^{v*}=0,\\ = \rho_{3k}^*, & \text{if } q_{jk}^{v*}>0.\end{cases} \tag{6.10b}$$

Conditions (6.10a) state that, if the expected demand price at demand market k is positive, then the quantities purchased by consumers at the demand market from the retailers in the aggregate is equal to the demand at demand market k. Conditions (6.10b) state, in turn, that in equilibrium, if the consumers at demand market k purchase the product from retailer j via transportation mode v, then the price charged by the retailer for the product plus the unit transaction cost is equal to the price that the consumers are willing to pay for the product. If the price plus the unit transaction cost exceeds the price the consumers are willing to pay at the demand market then there will be no transaction between the retailer and demand market via that transportation mode.

Equilibrium conditions (6.10a) and (6.10b) are equivalent to the following variational inequality problem, after summing over all demand markets: determine

$\left(Q^{2*},\rho_3^*\right) \in R_+^{noh+o}$satisfying:

$$\sum_{k=1}^{o}\left(\sum_{j=1}^{n}\sum_{v=1}^{h} q_{jk}^{v*} - \hat{d}_k\left(\rho_3^*\right)\right) \times \left[\rho_{3k} - \rho_{3k}^*\right]$$

$$+\sum_{k=1}^{o}\sum_{j=1}^{n}\sum_{v=1}^{h}\left(\rho_{2jk}^{v*} +_{jk}^{v}\left(Q^{2*}\right) - \rho_{3k}^*\right) \times \left[q_{jk}^{v} - q_{jk}^{v*}\right] \geq 0,$$

$$\forall \rho_3 \in R_+^o, \quad \forall Q^2 \in R_+^{noh}, \tag{6.11}$$

where ρ_3 is the o-dimensional vector with components: $\rho_{31}, \ldots, \rho_{3o}$ and Q^2 is the noh-dimensional vector.

Remark: In this chapter, we are interested in the cases where the expected demands are positive, that is, $\hat{d}_k(\rho_3) > 0, \forall \rho_3 \in R_+^o$ for $k = 1, \ldots, o$. Furthermore, we assume that the unit transaction costs: $c_{jk}^v\left(Q^2\right) > 0, \forall j, k, \forall Q^2 \neq 0$.

Under the above assumptions, we have that $\rho_{3k}^* > 0$ and $\hat{d}_k\left(\rho_3^*\right) = \sum_{j=1}^n \sum_{k=1}^o \sum_{v=1}^h q_{jk}^{v*}, \forall k$. This can be shown by contradiction. If there exists a $\bar{k}$ where $\rho_{3\bar{k}}^* = 0$, then according to (6.10a) we have that $\sum_{j=1}^n \sum_{k=1}^o \sum_{v=1}^h q_{jk}^{v*} \geq \hat{d}_{\bar{k}}\left(\rho_3^*\right) > 0$. Hence, there exists at least a $\left(j/\bar{k}\right)$ pair such that $q_{j\bar{k}}^{v*} > 0$, which means that $c_{j\bar{k}}^v\left(Q^{2*}\right) > 0$ by assumption. From conditions (6.10b), we have that $\rho_{2j\bar{k}}^{v*} + c_{j\bar{k}}^v\left(Q^{2*}\right) = \rho_{3\bar{k}} > 0$, which leads to a contradiction.

6.2.4 The Equilibrium Conditions of the Supply Chain

In equilibrium, we must have that the optimality conditions for all manufacturers, as expressed by (6.4), the optimality conditions for all retailers, as expressed by (6.9), and as well as the equilibrium conditions for all the demand markets, as expressed by (6.9), must hold simultaneously (see also Nagurney, et al., 2005). Hence, the product shipments of the manufacturers with the retailers must be equal to the product shipments that retailers accept from the manufacturers. We now formally state the equilibrium conditions for the entire supply chain network as follows:

Definition 5.1: Supply Chain Network Equilibrium with Uncertainty and Random Demands

The equilibrium state of the supply chain network with disruption risks and random demands is one where the flows of the product between the tiers of the decision-makers coincide and the flows and prices satisfy the sum of conditions (6.4), (6.9), and (6.11).

The summation of inequalities (6.4), (6.9), and (6.11), after algebraic simplification, yields the following result (see also Nagurney, 1999, 2006a).

Theorem 5.1: Variational Inequality Formulation

A product shipment and price pattern $\left(Q^{1*}, Q^{2*}, \gamma^*, \rho_3^*\right) \in R_+^{mng+noh+n+o}$ *is an equilibrium pattern of the supply chain model according to Definition 1, if and only if it satisfies the variational inequality problem:*

$$\sum_{i=1}^{m}\sum_{j=1}^{n}\sum_{u=1}^{g}\left[\frac{\partial \hat{F}_i\left(Q^{1*}\right)}{\partial q_{ij}^u}+\frac{\partial \hat{C}_{ij}^u\left(q_{ij}^{u*}\right)}{\partial q_{ij}^u}+\theta_i\left(\frac{\partial VF_i\left(Q^{1*}\right)}{\partial q_{ij}^u}+\frac{\partial VC_{ij}^u\left(q_{ij}^{u*}\right)}{\partial q_{ij}^u}\right)\right.$$
$$\left.+\frac{\partial \hat{C}_j^1\left(Q^{1*},Q^{2*}\right)}{\partial q_{ij}^u}+\varpi_j\frac{\partial VC_j^1\left(Q^{1*},Q^{2*}\right)}{\partial q_{ij}^u}-\gamma_j^*\right]\times\left[q_{ij}^u-q_{ij}^{u*}\right]$$
$$\sum_{j=1}^{n}\sum_{k=1}^{o}\sum_{v=1}^{g}\left[\frac{\partial \hat{C}_j^1\left(Q^{1*},Q^{2*}\right)}{\partial q_{jk}^v}+\varpi_j\frac{\partial VC_j^1\left(Q^{1*},Q^{2*}\right)}{\partial q_{jk}^v}\right.$$
$$\left.+\gamma_j^*+c_{jk}^v\left(Q^{2*}\right)-\rho_{3k}^*\right]\times\left[q_{jk}^v-q_{jk}^{v*}\right]$$
$$+\sum_{j=1}^{n}\left[\sum_{i=1}^{m}\sum_{u=1}^{g}q_{ij}^{u*}-\sum_{k=1}^{o}\sum_{v=1}^{h}q_{jk}^{v*}\right]\times\left[\gamma_j-\gamma_j^*\right]+\sum_{k=1}^{o}\left(\sum_{j=1}^{n}\sum_{v=1}^{h}q_{jk}^{v*}\right.$$
$$\left.-\hat{d}_k\left(\rho_3^*\right)\right)\times\left[\rho_{3k}-\rho_{3k}^*\right]\geq 0,$$
$$\forall\left(Q^1,Q^2,\gamma,\rho_3\right)\in R_+^{mng+noh+n+o}. \tag{6.12}$$

For easy reference in the subsequent sections, variational inequality problem (6.12) can be rewritten in standard variational inequality form (cf. Nagurney, 1999) as follows: determine $X^* \in K$:

$$\left\langle F\left(X^*\right), X-X^*\right\rangle \geq 0, \quad \forall X \in K \equiv R_+^{mng+noh+n+o}, \tag{6.13}$$

where

$$X \equiv \left(Q^1, Q^2, \gamma, \rho_3\right),$$
$$F(X) \equiv \left(F_{iju}, F_{jkv}, F_j, F_k\right)_{i=1,\dots,m;\,j=1,\dots,n;\,k=1,\dots,o;\,u=1,\dots,g;\,v=1,\dots,h},$$

and the specific components of $F$are given by the functional terms preceding the multiplication signs in (6.12). The term $\langle\cdot,\cdot\rangle$ denotes the inner product in N-dimensional Euclidean space.

Note that the equilibrium values of the variables in the model (which can be determined from the solution of either variational inequality (6.12) or (6.13)) are: the equilibrium product shipments between manufacturers and the retailers given by Q^{1*}, and the equilibrium product shipments transacted between the retailers and the demand markets given by Q^{2*}, as well as the equilibrium prices: ρ_3^* and γ^*. We now discuss how to recover the prices ρ_1^* associated with the top tier of nodes of the supply chain network and the prices ρ_2^* associated with the middle tier.

First, note that, from (6.5), we have that if $q_{ij}^{u*} > 0$, then price $\rho_{1ij}^{u*} = \frac{\partial \hat{F}_i(Q^{1*})}{\partial q_{ij}^u} + \frac{\partial \hat{C}_{ij}^u(q_{ij}^{u*})}{\partial q_{ij}^u} + \theta_i \left(\frac{\partial VF_i(Q^{1*})}{\partial q_{ij}^{u*}} + \frac{\partial VC_{ij}^u(q_{ij}^{u*})}{\partial q_{ij}^{u*}} \right)$. On the other hand, from (6.9), it follows that, if $q_{ij}^{v*} > 0$, the price $\rho_{2j}^* = \gamma_j^* + \frac{\partial \hat{C}_j^1(Q^{1*},Q^{2*})}{\partial q_{jk}^v} + \varpi_j \frac{\partial VC_j^1(Q^{1*},Q^{2*})}{\partial q_{jk}^v}$. These expressions can be utilized to obtain all such prices for all modes and decision-makers.

6.3 A Weighted Supply Chain Performance Measure

In this section, we first propose a supply chain network performance measure. Then, we provide the definition of supply chain network robustness, and follow with the definition for a weighted supply chain performance measure.

6.3.1 A Supply Chain Network Performance Measure

Recently, Qiang and Nagurney (2008) (see also Nagurney and Qiang, 2007a,b,c) proposed a network performance measure, which captures flows, costs, and behavior under network equilibrium conditions. Based on the measure in the above paper(s), we propose the following definition of a supply chain network performance measure.

Definition 5.2: The Supply Chain Network Performance Measure

The supply chain network performance measure,ε, for a given supply chain, and expected demands: $\hat{d}_k; k = 1, 2, \ldots, o$, is defined as follows:

$$\varepsilon \equiv \frac{\sum_{k=1}^{o} \frac{\hat{d}_k}{\rho_{3k}}}{o}, \tag{6.14}$$

where o is the number of demand markets in the supply chain network, and $\hat{d}_k$ and ρ_{3k} denote, respectively, the expected equilibrium demand and the equilibrium price at demand market k.

Note that the equilibrium price is equal to the unit production and transaction costs plus the weighted marginal risks for producing and transacting one unit from the manufacturers to the demand markets (see also Nagurney, 2006b). According to the above performance measure, a supply chain network performs well in network equilibrium if, on the average, and across all demand markets, a large demand can be satisfied at a low price. Therefore, in this chapter, we apply the above performance measure to assess the robustness of particular supply chain networks. From the discussion in Sect. 5.2.3, we have that $\rho_{3k} > 0, \forall k$. Therefore, the above definition is well-defined.

Furthermore, since each individual may have different opinions as to the risks, we need a "basis" to compare supply chain performance under different risk attitudes and to understand how risk attitudes affect the performance of a supply chain. Hence, we defineε^0as the supply chain performance measure where the $\hat{d}_k$and the ρ_{3k}; $k = 1, \ldots, o$, are obtained by assuming that the weights that reflect the manufacturers and the retailers' attitudes towards the disruption risks are zero. This definition excludes individuals' subjective differences in a supply chain and, with this definition, we are ready to study supply chain network robustness.

6.3.2 Supply Chain Robustness Measurement

Robustness has a broad meaning and is often couched in different settings. Generally speaking, robustness means that the system performs well when exposed to uncertain future conditions and perturbations (cf. Bundschuh et al., 2003; Snyder, 2003; Holmgren 2007).

Therefore, we propose the following rationale to assess the robustness of a supply chain: assume that all the random parameters take on a given threshold probability value; say, for example, 95 %. Moreover, assume that all the cumulative distribution functions for random parameters have inverse functions. Hence, we have that: $\alpha_i = F_i^{-1}(0.95)$, for $i = 1, \ldots, m$; $\beta_{ij}^u = F_{ij}^{u-1}(0.95)$, for $i = 1, \ldots, m$; $j = 1, \ldots, n$, and so on. With the disruption related parameters given, we can calculate the supply chain performance measure according to the definition given by (6.14). Let $\varepsilon_w$denote the supply chain performance measure with random parameters fixed at a certain level as described above. For example, when $w = 0.95$, ε_w is the supply chain performance with all the random risk parameters fixed at the value of a 95 % probability level. Then, the supply chain network robustness measure, R, is given by the following:

$$R = \varepsilon^0 - \varepsilon_w , \tag{6.15}$$

where ε^0 gauges the supply chain performance based on the model introduced in Sect. 6.2, but with weights related to risks being zero.

ε^0 examines the "base" supply chain performance while ε_w assesses the supply chain performance measure at some pre-specified uncertainty level. If their difference is small, a supply chain maintains its functionality well and we consider the supply chain to be robust at the threshold disruption level. Hence, the lower the value of R, the more robust a supply chain is. Note that since the random parameters are fixed at certain threshold level when we compute ε_w, the corresponding cost variances are equal to zero. Therefore, ε_w does not consider individual's risk attitude either as in the definition of ε^0.

Notably, the above robustness definition has implications for network resilience as well. *Resilience* is a general and conceptual term, which is hard to quantify.

McCarthy (2007) defined resilience "... as the ability of a system to recover from adversity, either back to its original state or an adjusted state based on new requirements, ...". For a comprehensive discussion of resilience, please refer to the Critical Infrastructure Protection Program (2007). Because our supply chain measure is based on the network equilibrium model, a network that is qualified as being robust according to our measure is also resilient provided that its performance after experiencing the disruption(s) is close to the "original value." Interestingly, this idea is in agreement with Hansson and Helgesson (2003), who proposed that robustness can be treated as a special case of resilience.

6.3.2.1 A Weighted Supply Chain Performance Measure

Note that different supply chains may have different requirements regarding the performance and robustness concepts introduced in the previous sections. For example, in the case of a supply chain of a toy product one may focus on how to satisfy demand in the most cost efficient way and not care too much about supply chain robustness. A medical/healthcare supply chain, on the other hand, may have a requirement that the supply chain be highly robust when faced with uncertain conditions. Hence, in order to be able to examine and to evaluate the different application-based supply chains from both perspectives, we now define a weighted supply chain performance measure as follows:

$$\hat{\varepsilon} = (1- \in)\, \varepsilon^0 + \in (-R) \;, \qquad (6.16)$$

where $\varepsilon \in [0, 1]$ is the weight that is placed on the supply chain robustness.

When ε is equal to 1, the performance of a supply chain hinges only on the robustness measure, which may be the case for a medical/healthcare supply chain, noted above. In contrast, when ε is equal to 0, the performance of the supply chain depends solely on how well it can satisfy demands at low prices. The supply chain of a toy product in the above discussion falls into this category.

6.4 Examples

The supply chain network topology for the numerical examples is depicted in Fig. 6.2 below. There are assumed to be two manufacturers, two retailers, and two demand markets. There are two modes of transportation available between each manufacturer and retailer pair and between each retailer and demand market pair. These examples are solved by the modified projection method of Korpelevich (1977); see also, e. g., Nagurney (2006a). Furthermore, for completeness, in the following examples, I also reported different variance functions for the general purpose of computing supply chain equilibrium solutions though they are not necessarily needed for determining the supply chain performance.

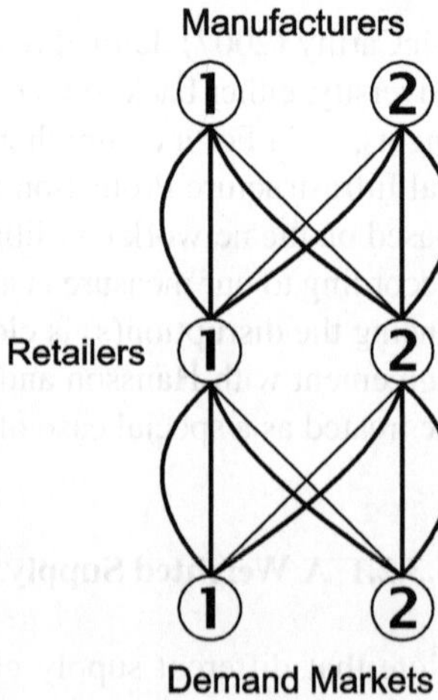

Fig. 6.2 The supply chain network for the numerical examples

Example 6.1

In the first example, for illustration purposes, we assumed that all the random parameters followed uniform distributions. The relevant parameters are as follows:

$$\alpha_i \sim [0, 2] \quad \text{for } i = 1, 2;$$
$$\beta_{ij}^u \sim [0, 1] \quad \text{for } i = 1, 2; \quad j = 1, 2; \quad u = 1, 2;$$
$$\eta_j \sim [0, 3] \quad \text{for } j = 1, 2\,.$$

We further assumed that the demand functions followed a uniform distribution given by $[200 - 2\rho_{3k}, 600 - 2\rho_{3k}]$, for $k = 1, 2$. Hence, the expected demand functions are:

$$\hat{d}_k(\rho_3) = 400 - 2\rho_{3k}, \text{ for } k = 1, 2\,.$$

The production cost functions for the manufacturers are given by:

$$f_1(Q^1, \alpha_1) = 2.5\left(\sum_{j=1}^{2}\sum_{u=1}^{2} q_{1j}^u\right)^2 + \left(\sum_{j=1}^{2}\sum_{u=1}^{2} q_{1j}^u\right)\left(\sum_{j=1}^{2}\sum_{u=1}^{2} q_{2j}^u\right) + 2\alpha_1\left(\sum_{j=1}^{2}\sum_{u=1}^{2} q_{1j}^u\right),$$

$$f_2(Q^1, \alpha_2) = 2.5\left(\sum_{j=1}^{2}\sum_{u=1}^{2} q_{2j}^u\right)^2 + \left(\sum_{j=1}^{2}\sum_{u=1}^{2} q_{1j}^u\right)\left(\sum_{j=1}^{2}\sum_{u=1}^{2} q_{2j}^u\right) + 2\alpha_2\left(\sum_{j=1}^{2}\sum_{u=1}^{2} q_{2j}^u\right).$$

The expected production cost functions for the manufacturers are given by:

$$\hat{F}_1\left(Q^1\right) = 2.5\left(\sum_{j=1}^{2}\sum_{u=1}^{2} q_{1j}^u\right)^2 + \left(\sum_{j=1}^{2}\sum_{u=1}^{2} q_{1j}^u\right)\left(\sum_{j=1}^{2}\sum_{u=1}^{2} q_{2j}^u\right) + 2\left(\sum_{j=1}^{2}\sum_{u=1}^{2} q_{1j}^u\right),$$

$$\hat{F}_2(Q^1) = 2.5\left(\sum_{j=1}^{2}\sum_{u=1}^{2} q_{2j}^u\right)^2 + \left(\sum_{j=1}^{2}\sum_{u=1}^{2} q_{1j}^u\right)\left(\sum_{j=1}^{2}\sum_{u=1}^{2} q_{2j}^u\right) + 2\left(\sum_{j=1}^{2}\sum_{u=1}^{2} q_{2j}^u\right).$$

The variances of the production cost functions for the manufacturers are given by:

$$VF_1(Q^1) = \frac{4}{3}\left(\sum_{j=1}^{2}\sum_{u=1}^{2} q_{1j}^u\right)^2,$$

$$VF_2(Q^1) = \frac{4}{3}\left(\sum_{j=1}^{2}\sum_{u=1}^{2} q_{2j}^u\right)^2.$$

The transaction cost functions faced by the manufacturers and associated with transacting with the retailers are given by:

$$c_{ij}^1\left(q_{ij}^1, \beta_{ij}^1\right) = 0.5\left(q_{ij}^1\right)^2 + 3.5\beta_{ij}^1 q_{ij}^1, \quad \text{for } i = 1,2; j = 1,2,$$
$$c_{ij}^2\left(q_{ij}^2, \beta_{ij}^2\right) = \left(q_{ij}^2\right)^2 + 5.5\beta_{ij}^2 q_{ij}^2, \quad \text{for } i = 1,2; j = 1,2.$$

The expected transaction cost functions faced by the manufacturers and associated with transacting with the retailers are given by:

$$\hat{C}_{ij}^1\left(q_{ij}^1\right) = 0.5\left(q_{ij}^1\right)^2 + 1.75 q_{ij}^1, \quad \text{for } i = 1,2; j = 1,2,$$
$$\hat{C}_{ij}^2\left(q_{ij}^2\right) = 0.5\left(q_{ij}^2\right)^2 + 2.75 q_{ij}^2, \quad \text{for } i = 1,2; j = 1,2.$$

The variances of the transaction cost functions faced by the manufacturers and associated with transacting with the retailers are given by:

$$VC_{ij}^1\left(q_{ij}^1\right) = 1.0208\left(q_{ij}^1\right)^2, \quad \text{for } i = 1,2; j = 1,2,$$
$$VC_{ij}^2\left(q_{ij}^2\right) = 2.5208\left(q_{ij}^2\right)^2, \quad \text{for } i = 1,2; j = 1,2.$$

The handling costs of the retailers, in turn, are given by:

$$c_j\left(Q^1, Q^2, \eta_j\right) = 0.5\left(\sum_{i=1}^{2}\sum_{u=1}^{2} q_{ij}^u\right)^2 + \eta_j\left(\sum_{i=1}^{2}\sum_{u=1}^{2} q_{ij}^u\right), \quad \text{for } j = 1, 2.$$

The expected handling costs of the retailers are given by:

$$\hat{C}_j^1\left(Q^1, Q^2\right) = 0.5\left(\sum_{i=1}^{2}\sum_{u=1}^{2} q_{ij}^u\right)^2 + 1.5\left(\sum_{i=1}^{2}\sum_{u=1}^{2} q_{ij}^u\right), \quad \text{for } j = 1, 2.$$

The variance of the handling costs of the retailers are given by:

$$VC_j\left(Q^1, Q^2\right) = \frac{3}{4}\left(\sum_{i=1}^{2}\sum_{u=1}^{2} q_{ij}^u\right)^2, \quad \text{for } j = 1, 2.$$

The unit transaction costs from the retailers to the demand markets are given by:

$$c_{jk}^1\left(Q^2\right) = 0.3q_{jk}^1, \quad \text{for } j = 1, 2; k = 1, 2,$$
$$c_{jk}^2\left(Q^2\right) = 0.6q_{jk}^2, \quad \text{for } j = 1, 2; k = 1, 2.$$

We assumed that the manufacturers and the retailers placed zero weights on the disruption risks as discussed in Sect. 6.3.1 to compute ε^0.

In the equilibrium, under the expected costs and demands, we have that the equilibrium shipments between manufacturers and retailers are: $q_{ij}^{1*} = 8.5022$, for $i = 1, 2$; $j = 1, 2$; $q_{ij}^{2*} = 3.7511$, for $i = 1, 2$; $j = 1, 2$; whereas the equilibrium shipments between the retailers and the demand markets are: $q_{jk}^{1*} = 8.1767$, for $j = 1, 2$; $k = 1, 2$; $q_{jk}^{2*} = 4.0767$, for $j = 1, 2$; $k = 1, 2$. Finally, the equilibrium prices are: $\rho_{31}^* = \rho_{32}^* = 187.7466$ and the expected equilibrium demands are: $\hat{d}_1 = \hat{d}_2 = 24.5068$. The supply chain performance measure is equal to $\varepsilon^0 = 0.1305$. Now, assume that $w = 0.95$; that is, all the random cost parameters are fixed at a 95 % probability level. The resulting supply chain performance measure is computed as $\varepsilon_w = 0.1270$. If we let $\varepsilon = 0.5$ (cf. (6.12)), which means that we place equal emphasis on performance and robustness of the supply chain, the weighted supply chain performance measure is $\hat{\varepsilon} = 0.0635$.

Example 5.2

For the same network structure and cost and demand functions, we now assume that the relevant parameters are changed as follows: $\alpha_i \sim [0, 4]$ for $i = 1, 2$; $\beta_{ij}^u \sim [0, 2]$ for $i = 1, 2$; $j = 1, 2$; $u = 1, 2$; $\eta_j \sim [0, 6]$ for $j = 1, 2$.

In the equilibrium, under the expected costs and demands, we have that the equilibrium shipments between manufacturers and retailers are now: $q_{ij}^{1*} = 8.6008$, for $i = 1, 2$; $j = 1, 2$; $q_{ij}^{2*} = 3.3004$, for $i = 1, 2$; $j = 1, 2$; whereas the equilibrium shipments between the retailers and the demand markets are: $q_{jk}^{1*} = 7.9385$, for $j = 1, 2$; $k = 1, 2$; $q_{jk}^{2*} = 3.9652$, for $j = 1, 2$; $k = 1, 2$. Finally, the equilibrium prices are: $\rho_{31}^* = \rho_{32}^* = 188.0963$ and the expected equilibrium demands are: $\hat{d}_1 = \hat{d}_2 = 23.8074$. The supply chain performance measure is equal to $\varepsilon^0 = 0.1266$. Similar to the above example, let us assume that $w = 0.95$; that is, all the random cost parameters are fixed at a 95 % probability level. The resulting supply chain performance measure is now: $\varepsilon_w = 0.1194$. If we let $\varepsilon = 0.5$, the weighted supply chain performance measure is $\hat{\varepsilon} = 0.0597$.

Observe that first example leads to a better measure of performance since the uncertain parameters do not have as great of an impact as in the second one for the cost functions under the given threshold level.

6.5 Summary and Conclusions

In this chapter, we developed a novel supply chain network model to study the demand-side as well as the supply-side risks, with the demand being random and the supply-side risks modeled as uncertain parameters in the underlying cost functions. This supply chain model generalizes several existing models by including multiple transportation modes from the manufacturers to the retailers, and from the retailers to the demand markets. We also proposed a weighted supply chain performance and robustness measure based on our recently derived network performance/efficiency measure and illustrated the supply chain network model through numerical examples for which the equilibrium prices and product shipments were computed and robustness analyses conducted. For future research, we plan on constructing further comprehensive metrics in order to evaluate supply chain network performance and to also apply the results in this chapter to empirically-based supply chain networks in different industries.

Acknowledgements The research of the first two authors was supported by the John F. Smith Memorial Fund at the Isenberg School of Management. This support is gratefully appreciated.

The authors would like to thank Professor Teresa Wu of Arizona State University and Professor Jennifer Blackhurst of Iowa State University for the invitation to contribute this chapter to the volume **Managing Supply Chain Risk and Vulnerability: Tools and Methods for Supply Chain Decision Makers**, Springer.

References

1. Bazaraa MS, Sherali, HD, Shetty, CM (1993) Nonlinear programming: Theory and algorithms, John Wiley & Sons, New York
2. Beamon MB (1998) Supply chain design and analysis: models and methods. Int. J. Production Economics 55:281–294

3. Beamon MB (1999) Measuring supply chain performance. Int. J. Production Economics 19:275–292
4. Bundschuh M, Klabjan D, Thurston DL (2003) Modeling robust and reliable supply chains. Optimization Online e-print. www.optimization-online.org
5. Canadian Competition Bureau (2006). Competition bureau concludes examination into gasoline price spike following hurricane Katrina. http://www.competitionbureau.gc.ca
6. Craighead CW, Blackhurst J, Rungtusanatham MJ et al. (2007). The severity of supply chain disruptions: Design characteristics and mitigation capabilities. Decis. Sci. 38:131–156
7. Dong J, Zhang D, Nagurney A (2002) Supply chain networks with multi-criteria decision-makers. In: Taylor MAP (ed) Transportation and traffic theory in the 21st century, Pergamon.
8. Dong J, Zhang D, Yan H, Nagurney A (2005). Multitiered supply chain networks: Multicriteria decision-making under uncertainty. Ann. Operations Res. 135:155–178
9. Fairtrade Foundation (2002) Spilling the bean. http://www.fairtrade.org.uk
10. Gupta D (1996) The (Q, r) inventory system with an unreliable supplier. INFOR 34:59–76
11. Hansson SO, Helgesson G (2003) What is stability? Synthese 136:219–235
12. Hendricks KB, Singhal VR (2005) An empirical analysis of the effect of supply chain disruptions on long-term stock price performance and risk of the firm. Production and Operations Management 14:35–52
13. Holmgren AJ (2007) A framework for vulnerability assessment of electric power systems. In: Murray A, Grubesic T (ed.) Critical infrastructure – reliability and vulnerability. Springer, New York
14. Kleindorfer PR, Saad GH (2005) Managing disruption risks in supply chains. Production and Operations Management 14:53–68
15. Korpelevich GM (1977) The extragradient method for finding saddle points and other problems. Matekon 13:35–49
16. Lai KH, Ngai EWT, Cheng TCE (2002) Measures for evaluating supply chain performance in transport logistics. Transportation Res. E 38:439–456
17. Lambert DM, Pohlen TL (2001) Supply chain metrics. Int. J. of Logistics Management 12:1–19
18. Latour A (2001) Trial by fire: a blaze in Albuquerque sets off major crisis for cell-phone giants. Wall Str. J., January 29
19. Lee H, Whang S (1999) Decentralized multi-echelon supply chains: Incentives and information. Management Sci. 45:633–640
20. McCarthy JA (2007) From protection to resilience: injecting "Moxie" into the infrastructure security continuum. Critical infrastructure protection program, Discussion paper series, George Mason University, Fairfax, Virginia
21. Nagurney A (1999) Network economics: A variational inequality approach, 2nd edn. Kluwer Academic Publishers, Dordrecht, The Netherlands
22. Nagurney A (2006a) Supply chain network economics: dynamics of prices, flows, and profits, Edward Elgar Publishing, Cheltenham, England
23. Nagurney A (2006b). On the relationship between supply chin and transportation network equilibria: A supernetwork equivalence with computations, Transportation Res. E 42:293–316
24. Nagurney A, Cruz J, Dong J et al. (2005) Supply chain networks, electronic commerce, and supply side and demand side risk. European J Operational Res. 164:120–142
25. Nagurney A, Dong J, Zhang D (2002) A supply chain network equilibrium model. Transportation Res. E 38:281–303
26. Nagurney A, Qiang Q (2007a) A transportation network efficiency measure that captures flows, behavior, and costs with applications to network component importance identification and vulnerability. Proc. of the POMS Conf., Dallas, Texas, May 4–7
27. Nagurney A, Qiang Q (2007b) A network efficiency measure with application to critical infrastructure networks. J. Glob. Optim. 40:261–275
28. Nagurney A, Qiang Q (2007c) A network efficiency measure for congested networks. Europhys. Lett. 79:38005

29. Nash JF (1950) Equilibrium points in n-person games. Proc. of the Natl. Academy of Sci. USA 36:48–49
30. Nash JF (1951) Noncooperative games. Ann. of Mathematics 54:286–298
31. Parlar M (1997) Continuous review inventory problem with random supply interruptions. European J. Operational Res. 99:366–385
32. Parlar M, Perry D (1996) Inventory models of future supply uncertainty with single and multiple suppliers. Nav. Res. Logistics 43:191–210
33. Qiang Q, Nagurney A (2008) A unified network performance measure with importance identification and the ranking of network components. Optim. Lett. 2:127–142
34. Santoso T, Ahmed S, Goetschalckx M et al. (2005) A stochastic programming approach for supply chain network design under uncertainty European J. Operational Res. 167:96–115
35. Sheffi Y (2005) The resilient enterprise: Overcoming vulnerability for competitive advantage, MIT Press, Cambridge, Massachusetts
36. Silberberg E, Suen W (2000) The structure of economics: A mathematics analysis, McGraw-Hill, New York
37. Snyder LV (2003) Supply chain robustness and reliability: models and algorithms. Ph.D. dissertation, Northwestern University, Department of Industrial Engineering and Management Sciences, Evanston, Illinois
38. Snyder LV, Shen ZJM (2006) Supply chain management under the threat of disruptions. The Bridge 36:39–45
39. Snyder LV, Daskin MS (2005). A reliability model for facility location: the expected failure cost case. Transportation Sci. 39:400–416
40. Tang CS (2006a) Robust strategies for mitigating supply chain disruptions. Int. J. Logistics 9:33–45
41. Tang CS (2006b) Perspectives in supply chain risk management. Int. J. Production Economics 103:451–488
42. Tomlin BT (2006) On the value of mitigation and contingency strategies for managing supply chain disruption risks. Management Sci. 52:639–657
43. Wagner SM, Bode C (2007) An empirical investigation into supply chain vulnerability. J. Purchasing and Supply Management 12:301–312
44. Wilson MC (2007) The impact of transportation disruptions on supply chain performance. Transportation Res. E 43:295–320
45. Wu T, Blackhurst J, Chidambaram V (2006) The model for inbound supply risk analysis. Computers in Industry 57:350–365

Chapter 7
The Effects of Network Relationships on Global Supply Chain Vulnerability

Jose M. Cruz

Abstract In this chapter, we analyze the effects of levels of social relationship on the global supply chain networks vulnerability. Relationship levels in our framework are assumed to influence transaction costs as well as risk for the decision-makers. We propose a network performance measure for the evaluation of the global supply chain networks efficiency and vulnerability. The measure captures risk, transaction cost, price, transaction flow, revenue, and demand information in the context of the decision-makers behavior the network. The network consists of manufacturers, retailers, and consumers. Manufacturers and retailers are multicriteria decision-makers who decide about their production and transaction quantities as well as the level of social relationship they want to pursue in order to maximize net return and minimize risk. The model allows us to investigate the interplay of the heterogeneous decision-makers in the supply chain and to compute the resultant equilibrium pattern of product outputs, transactions, product prices, and levels of social relationship. The results show that high levels of relationship can lead to lower overall cost and therefore lower price and higher product transaction. Moreover, we use the performance measure to assess which nodes in the supply networks are the most vulnerable in the sense that their removal will impact the performance of the network in the most significant way.

7.1 Introduction

In recent years, growing competition and emphasis on efficiency and cost reduction, as well as the satisfaction of consumer demands, have brought new challenges for businesses in the global marketplace. As a result companies are outsourcing and offshoring large portions of manufacturing, sourcing in low-cost countries, reducing inventories, streamlining the supply base, and collaborating more intensively with other supply chain actors [12, 19]. However, the increase in interfirm dependence as well as longer and more complex supply chain setups with globe-spanning operations have increased the vulnerability of supply chains to unexpected events [6, 17, 18, 36]. For example, recently, the threat of illness in the form of SARS (see [10]) has disrupted supply chains, as have terrorist threats (cf. [34]) and the natural disaster of Hurricane Katrina in 2005. Indeed, at the same time that supply chains have become increasingly globalized, their operation environment has become unpredictable and filled with uncertainty.

Supply chain disruptions can materialize from various areas internal and external to a supply chain. The main supply chain vulnerability drivers can be divided in three groups, demand-side, supply-side, and Catastrophic events. The demand-side drivers would include demand uncertainty, customer dependence, and disruptions in the physical distribution of products to the end-customers. The supply-side drivers include supplier business risks, production capacity constraints on the supply market, quality problems, technological changes, and product design changes [39]. Moreover these drivers would increase supply chain vulnerability even more if the firm uses single sourcing or fewer suppliers. Catastrophic events refer to natural hazards, socio-political instability, civil unrest, economic disruptions, and terrorist attacks [26, 24]. Therefore we believe that firms should proactively assess and manage the uncertainties in supply chain by creating a portfolio of relationships with their suppliers and demand markets in order to guard against costly supply chain disruptions.

The value of relationship is not only economical but also technical and social [14]. Strong supply chain relationships enable firms to react to changes in the market, create customer value and loyalty, which lead to improve profit margins [11]. The benefits are reduction of production, transportation and administrative costs. On the technical development the greatest benefit is the possibility of sharing the resources of suppliers and shortening the lead-times. Spekman and Davis [35] found that supply chain networks that exhibit collaborative behaviors tend to be more responsive and that supply chain-wide costs are, hence, reduced. These results are also supported by Dyer [9] who demonstrated empirically that a higher level of trust (relationship) lowers transaction costs (costs associated with negotiating, monitoring, and enforcing contracts). Baker and Faulkner [1] present an overview of papers by economic sociologists that show the important role of relationships due to their potential to reduce risk and uncertainty. Uzzi [37] and Gadde and Snehota [14] suggest that multiple relationships can help companies deal with the negative consequences related to dependence on supply chain partners. Krause et al. [25] found that buyer commitment and social capital accumulation with key suppliers can improve buying company performance. However, Christopher and Jüttner [5] indicate that the value of the relationship depends on the substitutability of the buyers or sellers, the indispensability of goods, savings resulting from partner's practices and the degree of common interest.

In this chapter, we analyze the effects of relationships on a multitiered global supply chain network efficiency and vulnerability. Wakolbinger and Nagurney [38] and Cruz et al. [8] developed a framework for the modeling and analysis of supply chains networks that included the role that relationships play. Their contribution was apparently the first to introduce relationship levels in terms of flows on networks, along with logistical flows in terms of product transactions, combined with pricing. However, their models did not considerer the effects of relationship levels on supply chain efficiency and vulnerability.

This chapter models the multicriteria decision-making behavior of the various decision-makers in a multitiered global supply chain network, which includes the maximization of profit and the minimization of risk through the inclusion of the social relationship, in the presence of both business-to-business (B2B) and business-to-consumer (B2C) transactions. We describe the role of relationships in the global supply chain networks. Decision-makers in a given tier of the network can decide on the relationship levels that they want to achieve with decision-makers associated with the other tiers of the network. Establishing/maintaining a certain relationship level induces some costs, but may also lower the risk and the transaction costs. We explicitly describe the role of relationships in influencing transaction costs and risk. Both the risk functions and the relationship cost functions are allowed to depend on the relationship levels. In addition, we analyze the effects of the levels of social relationship on the supply chain efficiency and vulnerability. Hence, we truly capture the effects of networks of relationships in the global supply chain framework.

This chapter is organized as follows. In Sect. 7.2, we develop the multitiered, multiperiod supply chain network model. We describe decision-makers' optimizing behavior and establish the governing equilibrium conditions along with the corresponding variational inequality formulation. In Sect. 7.3, we present the supply chain network efficiency and vulnerability measures. In Sect. 7.4, we present numerical examples. Section 7.5 provides managerial implications. We conclude the chapter with Sect. 7.6 in which we summarize our results and suggest directions for future research.

7.2 The Global Supply Chain Networks Model

In this Section, we develop the network model with manufacturers, retailers, and demand markets in a global context. We assume that the manufacturers are involved in the production of a homogeneous product and we consider L countries, with I manufacturers in each country, and J retailers, which are not country-specific but, rather, can be either physical or virtual, as in the case of electronic commerce. There are K demand markets for the homogeneous product in each country and H currencies in the global economy. We denote a typical country by l or $\hat{l}$, a typical manufacturer by i, and a typical retailer by j. A typical demand market, on the other hand, is denoted by k and a typical currency by h. We assume that each manufacturer can conduct transactions with the retailers in different currencies. The demand for the product in a country can be associated with a particular currency.

The network in Fig. 7.1 represents the global supply chain network consisting of three tiers of decision-makers. The top tier of nodes consists of the manufacturers in the different countries, with manufacturer i in country l being referred to as

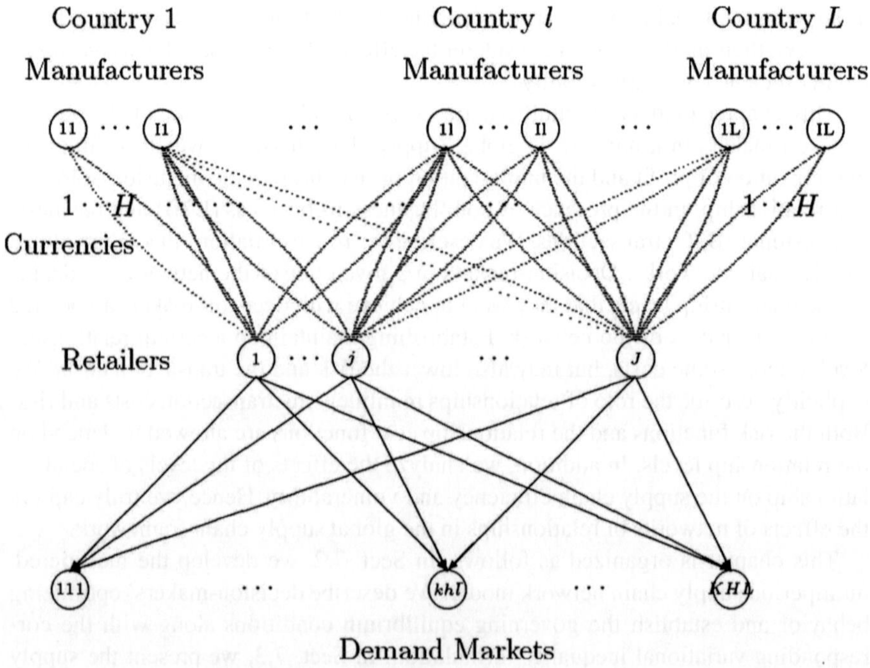

Fig. 7.1 The structure of the global supply chain network

manufacturer il and associated with node il. There are, hence, IL top-tiered nodes in the network. The middle tier of nodes of the network consists of the retailers (which recall need not be country-specific) and who act as intermediaries between the manufacturers and the demand markets, with a typical retailer j associated with node j in this (second) tier of nodes. The bottom tier in the supply chain network consists of the demand markets, with a typical demand market k in currency h and country $\hat{l}$, being associated with node $kh\hat{l}$ in the bottom tier of nodes. There are, as depicted in Fig. 7.1, J middle (or second) tiered nodes corresponding to the retailers and KHL bottom (or third) tiered nodes in the global supply chain network.

We have identified the nodes in the network and now we turn to the identification of the links joining the nodes in a given tier with those in the next tier. We assume that each manufacturer i in country l involved in the production of the homogeneous product can transact with a given retailer in any of the H available currencies, as represented by the H links joining each top tier node with each middle tier node j; $j = 1, \ldots, J$. Furthermore, each retailer (intermediate) node j; $j = 1, \ldots, J$, can transact with each demand market denoted by node $kh\hat{l}$. The product transactions represent the flows on the links of the supply chain network in Fig. 7.1.

We also assume that each manufacturer i in country l can establish a portfolio of relationships with retailers. Furthermore, each retailer (intermediate) node j; $j = 1, \ldots, J$, can establish a relationship level with a demand market denoted by node $kh\hat{l}$ and with manufacturers. We assume that the relationship levels are non-negative and that they may attain a value from 0 through 1. A relationship level of 0 indicates no relationship and a relationship of 1 indicates the highest possible relationship. These relationship levels are associated with each of nodes of the first two tiers of the network in Fig. 7.1.

Note that there will be prices associated with each of the tiers of nodes in the global supply chain network. The model also includes the rate of appreciation of currency h against the basic currency, which is denoted by e_h (see [28]). These "exchange" rates are grouped into the column vector $e \in R^H$. The variables for this model are given in Table 7.1. All vectors are assumed to be column vectors.

Table 7.1 Variables in global supply chain networks

Notation	Definition
q	IL-dimensional vector of the amounts of the product produced by the manufacturers in the countries with component il denoted by q^{il}
Q^1	$ILJH$-dimensional vector of the amounts of the product transacted between the manufacturers in the countries in the currencies with the retailers with component ${}^{il}_{jh}$ denoted by q^{il}_{jh}
Q^2	$JKHL$-dimensional vector of the amounts of the product transacted between the retailers and the demand markets in the countries and currencies with component ${}^{j}_{kh\hat{l}}$ denoted by $q^{j}_{kh\hat{l}}$
η^1	IL-dimensional vector of the relationships levels of manufacturers with component ^{il} denoted by η^{il}
η^2	J-dimensional vector of the relationship levels of retailers with component ^{j} denoted by η^j
ρ^{il}_{1jh}	Price associated with the product transacted between manufacturer il and retailer j in currency h
$\rho^{j}_{2kh\hat{l}}$	Price associated with the product transacted between retailer j and demand market k in currency h and country $\hat{l}$
$\rho3$	KHL-dimensional vector of the demand market prices of the product at the demand markets in the currencies and in the countries with component khl denoted by $\rho3khl$

We now turn to the description of the functions and assume that they are measured in the base currency (dollar). We first discuss the production cost, transaction cost, handling, and unit transaction cost functions given in Table 7.2. Each manufacturer is faced with a certain production cost function that may depend, in general, on the entire vector of production outputs. Furthermore, each manufacturer and each retailer are faced with transaction costs. The transaction costs are affected/influenced by the amount of the product transacted and the relationship levels. As indicated in the introduction, relationship levels affect transaction costs [9, 35]. This is especially

Table 7.2 Production, handling, transaction, and unit transaction cost functions

Notation	Definition
$f^{il}(q) = f^{il}(q^{il})$	The production cost function of manufacturer i in country l
$c_j(Q^1)$	The handling/conversion cost function of retailer j
$c^{il}_{jh}\left(q^{il}_{jh}, \eta^{il}\right)$	The transaction cost function of manufacturer il transacting with retailer j in currency h
$\hat{c}^{il}_{jh}\left(q^{il}_{jh}, \eta^{j}\right)$	The transaction cost function of retailer j transacting with manufacturer il in currency h
$c^{j}_{kh\hat{l}}\left(q^{j}_{kh\hat{l}}, \eta^{j}\right)$	The transaction cost function of retailer j transacting with demand market $kh\hat{l}$
$\hat{c}^{j}_{kh\hat{l}}(Q^2, \eta^2)$	The unit transaction cost function associated with consumers at demand market khl in obtaining the product from retailer j

important in international exchanges in which transaction costs may be significant. Hence, the transaction cost functions depend on flows and relationship levels.

Each retailer is also faced with what we term a handling/conversion cost (cf. Table 7.2), which may include, for example, the cost of handling and storing the product plus the cost associated with transacting in the different currencies. The handling/conversion cost of a retailer is a function of how much he has obtained of the product from the various manufacturers in the different countries and what currency the transactions took place. For the sake of generality, however, we allow the handling functions to depend also on the amounts of the product held and transacted by other retailers.

The consumers at each demand market are faced with a unit transaction cost. As in the case of the manufacturers and the retailers, higher relationship levels may potentially reduce transaction costs, which mean that they can lead to quantifiable cost reductions. The unit transaction costs depend on the amounts of the product that the retailers and the manufacturers transact with the demand markets as well as on the vectors of relationships established with the demand markets. The generality of the unit transaction cost function structure enables the modeling of competition on the demand side. Moreover, it allows for information exchange between the consumers at the demand markets who may inform one another as to their relationship levels which, in turn, can affect the transaction costs.

We now turn to the description of the relationship production cost functions and, finally, the risk functions and the demand functions. We assume that the relationship production cost functions as well as the risk functions are convex and continuously differentiable. The demand functions are assumed to be continuous.

We start by describing the relationship production cost functions that are given in Table 7.3. We assume that each manufacturer may actively try to achieve a certain relationship level with a retailer as proposed in Golicic et al. [15]. Furthermore, each retailer may actively try to achieve a certain relationship level with a manufacturer

Table 7.3 Relationship productions cost functions

Notation	Definition
$b^{il}\left(\eta^{il}\right)$	The relationship production cost function associated with manufacturer il
$b^{j}\left(\eta^{j}\right)$	The relationship production cost function associated with retailer j

and/or demand market. These relationship production cost functions may be distinct for each such combination. Their specific functional forms may be influenced by such factors as the willingness of retailers or demand markets to establish/maintain a relationship as well as the level of previous business relationships and private relationships that exist. Hence, we assume that these production cost functions are also affected and influenced by the relationship levels. Crosby and Stephens [7] indicate that the relationship strength changes with the amount of buyer-seller interaction and communication. In a global setting, cultural differences, difficulties with languages, and distances, may also play a role in making it more costly to establish (and to maintain) a specific relationship level (cf. [20]).

The concept of relationship levels was inspired by a paper by Golicic et al. [15] who introduced the concept of relationship magnitude. That research strongly suggested that different relationship magnitudes lead to different benefits and those different levels of relationship magnitudes can be achieved by putting more or less time and effort into the relationship. The idea of a continuum of relationship strength is also supported by several theories of relationship marketing that suggest that business relationships vary on a continuum from transactional to highly relational (cf. [13]). The model by Wakolbinger and Nagurney [38] operationalized the frequently mentioned need to create a portfolio of relationships (cf. [4, 15]). The optimal portfolio balanced out the various costs and the risk, against the profit and the relationship value and included the individual decision-makers preferences and risk aversions.

We now describe the risk functions as presented in Table 7.4. We note that the risk functions in our model are functions of both the product transactions and the relationship levels. Jüttner et al. [21] suggest that supply chain-relevant risk sources falls into three categories: environmental risk sources (e. g., fire, social-political actions, or "acts of God"), organizational risk sources (e. g., production uncertainties), and network-related risk sources. Johnson [22] and Norrman and Jansson [31] argue that network related risk arises from the interaction between organizations within the supply chain, e. g., due to insufficient interaction and cooperation. Here, we model

Table 7.4 Risk funktions

Notation	Definition
$r^{il}\left(Q^1, \eta^1\right)$	The risk incurred by manufacturer il in his transactions
$r^{j}\left(Q^1, Q^2, \eta^2\right)$	The risk incurred by retailer j in his transactions

supply chain organizational risk and network-related risk by defining the risk as a function of product flows as well as relationship levels. We use relationship levels (levels of cooperation) as a way of possibly mitigating network-related risk. We also note that by including the currency appreciation rate (e_h) in our model, in order to convert the prices to the base currency (dollar), we are actually mitigating any exchange rate risk. Of course, in certain situations; see also Granovetter [16], the risk may be adversely affected by higher levels of relationships. Nevertheless, the functions in Table 7.4 explicitly include relationship levels and product transactions as inputs into the risk functions and reflect this dependence.

The demand functions as given in Table 7.5 are associated with the bottom-tiered nodes of the global supply chain network. The demand of consumers for the product at a demand market in a currency and country depends, in general, not only on the price of the product at that demand market (and currency and country) but also on the prices of the product at the other demand markets (and in other countries and currencies). Consequently, consumers at a demand market, in a sense, also compete with consumers at other demand markets.

Table 7.5 Demand functions

Notation	Definition
$d_{kh\hat{l}}(\rho 3)$	The demand for the product at demand market k transacted in currency h in country $\hat{l}$ as a function of the demand market price vector

We now turn to describing the behavior of the various economic decision-makers. The model is presented, for ease of exposition, for the case of a single homogeneous product. It can also handle multiple products through a replication of the links and added notation. We first focus on the manufacturers. We then turn to the retailers, and, subsequently, to the consumers at the demand markets.

7.2.1 The Behavior of the Manufacturers

The manufacturers are involved in the production of a homogeneous product and in transacting with the retailers. Furthermore, they are also involved in establishing the corresponding relationship levels. The quantity of the product produced by manufacturer il must satisfy the following conservation of flow equation:

$$q^{il} = \sum_{j=1}^{J} \sum_{h=1}^{H} q^{il}_{jh}, \tag{7.1}$$

which states that the quantity of the product produced by manufacturer il is equal to the sum of the quantities transacted between the manufacturer and all retailers.

Hence, in view of (7.1), and as noted in Table 7.2, we have that for each manufacturer il the production cost $f^{il}(q) = f^{il}\left(q^{il}\right)$.

Each manufacturer il tries to maximize his profits. He faces total costs that equal the sum of his production cost plus the total transaction costs and the costs that he incurs in establishing and maintaining his relationship levels. His revenue, in turn, is equal to the sum of the price that he can obtain times the exchange rate multiplied by quantities of the product transacted. Furthermore, each manufacturer tries to minimize his risk generated by interacting with the retailers subject to his individual weight assignment to this criterion.

7.2.2 The Multicriteria Decision-Making Problem Faced by a Manufacturer

We can now construct the multicriteria decision-making problem facing a manufacturer which allows him to weight the criteria of profit maximization and risk minimization in an individual manner. Manufacturer il's multicriteria decision-making objective function is denoted by U^{il}. Assume that manufacturer il assigns a nonnegative weight α^{il} to the risk generated. The weight associated with profit maximization serves as the numeraire and is set equal to 1. The nonnegative weights measure the importance of the risk and, in addition, transform it value into monetary units. Let now ρ_{1jh}^{il*} denote the actual price charged by manufacturer il for the product in currency h to retailer j. We later discuss how such price is recovered. We can now construct a value function for each manufacturer (cf. [8, 23]) using a constant additive weight value function. Therefore, the multicriteria decision-making problem of manufacturer il can be expressed as:

$$\text{Maximize } U^{il} = \sum_{j=1}^{J}\sum_{h=1}^{H}(\rho_{1jh}^{il*} \times e_h)q_{jh}^{il} - f^{il}(q^{il}) - \sum_{j=1}^{J}\sum_{h=1}^{J} c_{jh}^{il}(q_{jh}^{il}, \eta^{il}) - b^{il}(\eta^{il}) - \alpha^{il} r^{il}(Q^1, \eta^1) \tag{7.2}$$

subject to:

$$q_{jh}^{il} \geq 0, \forall j, h, \tag{7.3}$$

$$0 \leq \eta^{il} \leq 1. \tag{7.4}$$

The first four terms on the right-hand side of the equal sign in (7.2) represent the profit which is to be maximized, the next term represents the weighted total risk which is to be minimized. The relationship values lie in the range between 0 and 1 and, hence, we need constraint (7.4).

7.2.3 The Optimality Conditions of Manufacturers

Here we assume that the manufacturers compete in a noncooperative fashion following Nash [29, 30]. Hence, each manufacturer seeks to determine his optimal strategies, that is, product transactions, given those of the other manufacturers. The optimality conditions of all manufacturers i; $i = 1, \ldots, I$; in all countries: l; $l = 1, \ldots, L$ simultaneously, under the above assumptions (cf. [2, 8, 27]), can be compactly expressed as: determine $(Q^{1*}, \eta^{1*}) \in k^1$, satisfying

$$\sum_{i=1}^{I}\sum_{l=1}^{L}\sum_{j=1}^{J}\sum_{h=1}^{H}\left[\frac{\partial f^{il}(q^{il*})}{\partial q^{il}_{jh}} + \frac{\partial c^{il}_{jh}(q^{il*}_{jh}, \eta^{il*})}{\partial q^{il}_{jh}} + \alpha^{il}\frac{\partial r^{il}(Q^{1*}, \eta^{1*})}{\partial q^{il}_{jh}} \right.$$
$$\left. -\rho^{il*}_{1jk} \times e_h\right] \times \left[q^{il}_{jh} - q^{il*}_{jh}\right]$$
$$+\sum_{i=1}^{I}\sum_{l=1}^{L}\sum_{j=1}^{J}\sum_{h=1}^{H}\left[\frac{\partial c^{il}_{jh}(q^{il*}_{jh}, \eta^{il*})}{\partial \eta^{il}} + \frac{\partial b^{il}(\eta^{il*})}{\partial \eta^{il}}\right.$$
$$\left.+\alpha^{il}\frac{\partial r^{il}(Q^{1*}, \eta^{1*})}{\partial \eta^{il}}\right] \times \left[\eta^{il} - \eta^{il*}\right] \geq 0, \quad \forall (Q^1, \eta^1) \in k^1, \tag{7.5}$$

where

$$k^1 \equiv \left[(Q^1, \eta^1) | q^{il}_{jh} \geq 0,\ 0 \leq \eta^{il} \leq 1, \quad \forall i, l, j, h\right]. \tag{7.6}$$

The inequality (7.5), which is a variational inequality (cf. [27]) has a meaningful economic interpretation. From the first term in (7.5) we can see that, if there is a positive volume of the product transacted from a manufacturer to a retailer, then the marginal cost of production plus the marginal cost of transacting plus the weighted marginal cost of risk must be equal to the price (times the exchange rate) that the retailer is willing to pay for the product. If that sum, in turn, exceeds that price then there will be no product transacted.

The second term in (7.5) show that if there is a positive relationship level (and that level is less than one) then the marginal cost associated with the level is equal to the marginal reduction in transaction costs plus the weighted marginal reduction in risk.

7.2.4 The Behavior of the Retailers

The retailers (cf. Fig. 7.1), in turn, are involved in transactions both with the manufacturers in the different countries, as well as with the ultimate consumers associated with the demand markets for the product in different countries and currencies and represented by the bottom tier of nodes in both the global supply chain network and the social network.

As in the case of manufacturers, the retailers have to bear some costs to establish and maintain relationship levels with manufacturers and with the consumers, who are the ultimate purchasers/buyers of the product. Furthermore, the retailers have associated transaction costs in regards to transacting with the manufacturers, which we assume can be dependent on the type of currency as well as the manufacturer. Retailers also are faced with risk in their transactions. As in the case of the manufacturers, the transaction cost functions and the risk functions depend on the amounts of the product transacted as well as the relationship levels.

Each retailer j tries to maximize profits and to minimize his individual risk associated with his transactions with these criteria weighted in an individual fashion.

7.2.5 A Retailer's Multicriteria Decision-Making Problem

Retailer j assigns a nonnegative weight δ^j to his risk. The weight associated with profit maximization is set equal to 1 and serves as the numeraire (as in the case of the manufacturers). The actual price charged for the product by retailer j is denoted by $\rho^{j*}_{2kh\hat{l}}$, and is associated with transacting with consumers at demand market k in currency h and country $\hat{l}$. Later, we discuss how such prices are arrived at. We are now ready to construct the multicriteria decision-making problem faced by a retailer which combines the individual weights with the criteria of profit maximization and risk minimization. Let U^j denote the multicriteria objective function associated with retailer j with his multicriteria decision-making problem expressed as:

Maximize

$$
\begin{aligned}
U^j = & \sum_{k=1}^{K}\sum_{h=1}^{H}\sum_{\hat{l}=1}^{L}(\rho^{j*}_{2kh\hat{l}} \times e_h) q^j_{kh\hat{l}} - c_j(Q^1) - \sum_{i=1}^{I}\sum_{l=1}^{L}\sum_{h=1}^{H}\hat{c}^{il}_{jh}(q^{il}_{jh}, \eta^j) \\
& - \sum_{k=1}^{K}\sum_{h=1}^{H}\sum_{\hat{l}=1}^{L} c^j_{kh\hat{l}}(q^j_{kh\hat{l}}, \eta^j) - b^j(\eta^j) - \sum_{i=1}^{I}\sum_{l=1}^{L}\sum_{h=1}^{H}(\rho^{il*}_{1jh} \times e_h) q^{il}_{jh} \\
& - \delta^j r^j(Q^1, Q^2, \eta^2)
\end{aligned}
\tag{7.7}
$$

subject to:

$$
\sum_{k=1}^{K}\sum_{h=1}^{H}\sum_{\hat{l}=1}^{L} q^j_{kh\hat{l}} \leq \sum_{i=1}^{I}\sum_{l=1}^{L}\sum_{h=1}^{H} q^{il}_{jh}, \tag{7.8}
$$

$$
q^{il}_{jh} \geq 0, \; q^j_{kh\hat{l}} \geq 0, \quad \forall i, l, k, h, \hat{l}, \tag{7.9}
$$

$$
0 \leq \eta^j \leq 1, \; \forall j. \tag{7.10}
$$

The first six terms on the right-hand side of the equal sign in (7.7) represent the profit which is to be maximized, the next term represents the weighted risk which is to be minimized. Constraint (7.8) states that consumers cannot purchase more of the product from a retailer than is held "in stock".

7.2.6 The Optimality Conditions of Retailers

We now turn to the optimality conditions of the retailers. Each retailer faces the multicriteria decision-making problem (7.7), subject to (7.8), the nonnegativity assumption on the variables (7.9), and the assumptions for the relationship values (7.10). As in the case of the manufacturers, we assume that the retailers compete in a noncooperative manner, given the actions of the other retailers. Retailers seek to determine the optimal transactions associated with the demand markets and with the manufacturers. In equilibrium, all the transactions between the tiers of the decision-makers will have to coincide, as we will see later in this section.

If one assumes that the handling, transaction cost, and risk functions are continuously differentiable and convex, then the optimality conditions for all the retailers satisfy the variational inequality: determine $(Q^{1*}, Q^{2*}, \eta^{2*}, \lambda^*) \in k^2$, such that

$$\sum_{j=1}^{J}\sum_{i=1}^{I}\sum_{l=1}^{L}\sum_{h=1}^{H}\left[\delta^j\frac{\partial r^{j*}}{\partial q_{jh}^{il}}+\frac{\partial c_j(Q^{1*})}{\partial q_{jh}^{il}}+\rho_{1jh}^{il*}\times e_h+\frac{\partial \hat{c}_{jh}^{il}(q_{jh}^{il*},\eta^{j*})}{\partial q_{jh}^{il}}-\lambda_j^*\right]$$
$$\times\left[q_{jh}^{il}-q_{jh}^{il*}\right]$$

$$+\sum_{j=1}^{J}\sum_{k=1}^{K}\sum_{h=1}^{H}\sum_{\hat{l}=1}^{L}\left[\delta^j\frac{\partial r^{j*}}{\partial q_{kh\hat{l}}^{j}}+\frac{\partial c_{kh\hat{l}}^{j}(q_{kh\hat{l}}^{j*},\eta^{j*})}{\partial q_{kh\hat{l}}^{j}}+\rho_{2kh\hat{l}}^{j*}\times e_h+\lambda_j^*\right]$$
$$\times\left[q_{kh\hat{l}}^{j}-q_{kh\hat{l}}^{j*}\right]$$

$$+\sum_{j=1}^{J}\sum_{k=1}^{K}\sum_{h=1}^{H}\sum_{\hat{l}=1}^{L}\left[\delta^j\frac{\partial r^{j*}}{\partial \eta^j}+\frac{\partial c_{kh\hat{l}}^{j}(q_{kh\hat{l}}^{j*},\eta^{j*})}{\partial \eta^j}+\frac{\partial \hat{c}_{jh}^{il}(q_{jh}^{il*},\eta^{j*})}{\partial \eta^j}\right.$$
$$\left.+\frac{\partial b^j(\eta^{j*})}{\partial \eta^j}\right]\times\left[\eta^j-\eta^{j*}\right]$$

$$+\sum_{j=1}^{J}\left[\sum_{i=1}^{I}\sum_{l=1}^{L}\sum_{h=1}^{H}q_{jh}^{il*}-\sum_{k=1}^{K}\sum_{h=1}^{H}\sum_{l=1}^{L}q_{kh\hat{l}}^{j*}\right]\times\left[\lambda_j-\lambda_j^*\right]\geq 0,$$

$$\forall (Q^1, Q^2, \eta^2, \lambda) \in k^2, \tag{7.11}$$

where $r^{j*} = r^j(Q^1, Q^2, \eta^{2*})$ and

$$k^2 \equiv \left[(Q^1, Q^2, \eta^2, \lambda)|q^{il}_{jh} \geq 0, q^j_{kh\hat{l}} \geq 0, 0 \leq \eta^j \leq 1, \lambda_j \geq 0, \quad \forall i, l, j, h, k, \hat{l}\right]. \tag{7.12}$$

Here λ_j denotes the Lagrange multiplier associated with constraint (7.8) and λ is the column vector of all the retailers' Lagrange multipliers. These Lagrange multipliers can also be interpreted as shadow prices. Indeed, according to the fourth term in (7.11), λ^*_j serves as the price to "clear the market" at retailer j.

The economic interpretation of the retailers' optimality conditions is very interesting. The first term in (7.11) states that if there is a positive amount of product transacted between a manufacturer/retailer pair and currency h, that is, $q^{il*}_{jh} > 0$, then the shadow price at the retailer, λ^*_j, is equal to the price charged for the product plus the various marginal costs and the associated weighted marginal risk. In addition, the second term in (7.11) shows that, if consumers at demand market $kh\hat{l}$ purchase the product from a particular retailer j, which means that, if the $q^{j*}_{kh\hat{l}m}$ is positive, then the price charged by retailer j, $\rho^{j*}_{2kh\hat{l}}$, is equal to λ^*_j plus the marginal transaction costs in dealing with the demand market and the weighted marginal costs for the risk that he has to bear. One also obtains interpretations from (7.11) as to the economic conditions at which the relationship levels associated with retailers interacting with either the manufacturers or the demand markets will take on positive values.

7.2.7 The Consumers at the Demand Markets

We now describe the consumers located at the demand markets. The consumers can transact through with the retailers. The consumers at demand market k in country $\hat{l}$ take into account the price charged for the product transacted in currency h by retailer j, which is denoted by $\rho^{j*}_{2kh\hat{l}}$ and the exchange rate, plus the transaction costs, in making their consumption decisions. The equilibrium conditions for demand market $kh\hat{l}$, thus, take the form: for all retailers: $j = 1, \ldots, J$, demand markets k; $k = 1, \ldots, K$; and currencies: h; $h = 1, \ldots, H$

$$\rho^{j*}_{2kh\hat{l}} \times e_h + \hat{c}^j_{kh\hat{l}}(Q^{2*}, \eta^{2*}) \begin{cases} = \rho^*_{3kh\hat{l}}, & \text{if } q^{j*}_{kh\hat{l}} > 0, \\ \geq \rho^*_{3kh\hat{l}}, & \text{if } q^{j*}_{kh\hat{l}} = 0, \end{cases} \tag{7.13}$$

In addition, we must have that for all $k, h, \hat{l}$

$$d_{kh\hat{l}}(\rho^*_3) \begin{cases} = \sum_{j=1}^{J} q^{j*}_{kh\hat{l}}, & \text{if } \rho^*_{3kh\hat{l}} > 0, \\ \leq \sum_{j=1}^{J} q^{j*}_{kh\hat{l}}, & \text{if } \rho^*_{3kh\hat{l}} = 0, \end{cases} \tag{7.14}$$

Conditions (7.13) state that consumers at demand market $kh\hat{l}$ will purchase the product from retailer j, if the price charged by the retailer for the product times the exchange rate plus the transaction cost (from the perspective of the consumer) does not exceed the price that the consumers are willing to pay for the product in that currency and country, i. e., $\rho^*_{3kh\hat{l}}$. Note that, according to (7.13), if the transaction costs are identically equal to zero, then the price faced by the consumers for a given product is the price charged by the retailer for the particular product.

Condition (7.14), on the other hand, states that, if the price the consumers are willing to pay for the product at a demand market is positive, then the quantity of the product at the demand market is precisely equal to the demand.

In equilibrium, conditions (7.13) and (7.14) will have to hold for all demand markets and these, in turn, can be expressed also as an inequality analogous to those in (7.5) and (7.11) and given by:
determine $(Q^{2*}, \rho_3^*) \in R_+^{(J+1)KHL}$, such that

$$\sum_{j=1}^{J}\sum_{k=1}^{K}\sum_{h=1}^{H}\sum_{\hat{l}=1}^{L}\left[\rho^{j*}_{2kh\hat{l}} \times e_h + \hat{c}^j_{kh\hat{l}}(Q^{2*}, \eta^{2*}) - \rho^*_{3kh\hat{l}}\right] \times \left[q^j_{kh\hat{l}} - q^{j*}_{kh\hat{l}}\right]$$

$$+\sum_{k=1}^{K}\sum_{h=1}^{H}\sum_{\hat{l}=1}^{L}\left[\sum_{j=1}^{J} q^{j*}_{kh\hat{l}} - d_{kh\hat{l}}(\rho_3^*)\right] \times \left[\rho_{3kh\hat{l}} - \rho^*_{3kh\hat{l}}\right] \geq 0,$$

$$\forall (Q^2, \rho_3) \in R_+^{(J+1)KHL}. \tag{7.15}$$

In the context of the consumption decisions, we have utilized demand functions, whereas profit functions, which correspond to objective functions, were used in the case of the manufacturers and the retailers. Since we can expect the number of consumers to be much greater than that of the manufacturers and retailers we believe that such a formulation is more natural. Also, note that the relationship levels in (7.15) are assumed as given. They are endogenous to the integrated model as is soon revealed.

7.2.8 The Equilibrium Conditions of the Network

In equilibrium, the product flows that the manufacturers in different countries transact with the retailers must coincide with those that the retailers actually accept from them. In addition, the amounts of the product that are obtained by the consumers in the different countries and currencies must be equal to the amounts that the retailers actually provide. Hence, although there may be competition between decision-makers at the same level of tier of nodes of the network there must be cooperation between decision-makers associated with pairs of nodes. Thus, in equilibrium, the prices and product transactions must satisfy the sum of the optimality conditions

(7.5) and (7.11) and the equilibrium conditions (7.15). We make these statements rigorous through the subsequent definition and variational inequality derivation.

Definition 1 (Network Equilibrium). The equilibrium state of the global supply chain network is one where the product transactions and relationship levels between the tiers of the network coincide and the product transactions, relationship levels, and prices satisfy the sum of conditions (7.5), (7.11), and (7.15).

The equilibrium state is equivalent to the following:

Theorem 1 (Variational Inequality Formulation). The equilibrium conditions governing the global supply chain network model according to Definition 1 are equivalent to the solution of the variational inequality given by:
determine $(Q^{1*}, Q^{2*}, \eta^{1*}, \eta^{2*}, \lambda^*, \rho_3^*) \in k$, satisfying:

$$\sum_{i=1}^{I}\sum_{l=1}^{L}\sum_{j=1}^{J}\sum_{h=1}^{H}\left[\frac{\partial f^{il}(q^{il*})}{\partial q^{il}_{jh}}+\frac{\partial c^{il}_{jh}(q^{il*}_{jh},\eta^{il*})}{\partial q^{il}_{jh}}+\frac{\partial \hat{c}^{il}_{jh}(q^{il*}_{jh},\eta^{j*})}{\partial q^{il}_{jh}}+\frac{\partial c_j(Q^{1*})}{\partial q^{il}_{jh}}\right.$$

$$\left.+\alpha^{il}\frac{\partial r^{il}(Q^{1*},\eta^{1*})}{\partial q^{il}_{jh}}+\delta^j\frac{\partial r^j(Q^{1*},Q^{2*},\eta^{2*})}{\partial q^{il}_{jh}}-\lambda^*_j\right]\times\left[q^{il}_{jh}-q^{il*}_{jh}\right]$$

$$+\sum_{j=1}^{J}\sum_{k=1}^{K}\sum_{h=1}^{H}\sum_{\hat{l}=1}^{L}\left[\delta^j\frac{\partial r^j(Q^{1*},Q^{2*},\eta^{2*})}{\partial q^{j}_{kh\hat{l}}}+\frac{\partial c^{j}_{kh\hat{l}}(q^{j*}_{kh\hat{l}},\eta^{j*})}{\partial q^{j}_{kh\hat{l}}}\right.$$

$$\left.+\ \hat{c}^{j}_{kh\hat{l}}(Q^{2*},\eta^{2*})+\lambda^*_j-\rho^*_{3kh\hat{l}}\right]\times\left[q^{j}_{kh\hat{l}}-q^{j*}_{kh\hat{l}}\right]$$

$$+\sum_{i=1}^{I}\sum_{l=1}^{L}\sum_{j=1}^{J}\sum_{h=1}^{H}\left[\frac{\partial c^{il}_{jh}(q^{il*}_{jh},\eta^{il*})}{\partial \eta^{il}}+\frac{\partial b^{il}(\eta^{il*})}{\partial \eta^{il}}+\alpha^{il}\frac{\partial r^{il}(Q^{1*},\eta^{1*})}{\partial \eta^{il}}\right]$$

$$\times\left[\eta^{il}-\eta^{il*}\right]+\sum_{j=1}^{J}\sum_{k=1}^{K}\sum_{h=1}^{H}\sum_{\hat{l}=1}^{L}\left[\delta^j\frac{\partial r^j(Q^{1*},Q^{2*},\eta^{2*})}{\partial \eta^j}+\frac{\partial c^{j}_{kh\hat{l}}(q^{j*}_{kh\hat{l}},\eta^{j*})}{\partial \eta^j}\right.$$

$$\left.+\frac{\partial \hat{c}^{il}_{jh}(q^{il*}_{jh},\eta^{j*})}{\partial \eta^j}+\frac{\partial b^j(\eta^{j*})}{\partial \eta^j}\right]\times\left[\eta^j-\eta^{j*}\right]+\sum_{j=1}^{J}\left[\sum_{i=1}^{I}\sum_{l=1}^{L}\sum_{h=1}^{H}q^{il*}_{jh}\right.$$

$$\left.-\sum_{k=1}^{K}\sum_{h=1}^{H}\sum_{\hat{l}=1}^{L}q^{j*}_{kh\hat{l}}\right]\times\left[\lambda_j-\lambda^*_j\right]+\sum_{k=1}^{K}\sum_{h=1}^{H}\sum_{\hat{l}=1}^{L}\left[\sum_{j=1}^{J}q^{j*}_{kh\hat{l}}-d_{kh\hat{l}}(\rho^*_3)\right]$$

$$\times\left[\rho_{3kh\hat{l}}-\rho^*_{3kh\hat{l}}\right]\geq 0,\quad \forall(Q^1,Q^2,\eta^1,\eta^2,\lambda,\rho_3)\in k, \tag{7.16}$$

where

$$k \equiv \Big[(Q^1, Q^2, \eta^1, \eta^2, \lambda, \rho_3) | q^{il}_{jh} \geq 0, q^{j}_{kh\hat{l}} \geq 0, 0 \leq \eta^{il} \leq 1\ ,$$
$$0 \leq \eta^j \leq 1, \rho_{3kh\hat{l}} \geq 0, \lambda_j \geq 0,\ \forall i, l, j, h, k, \hat{l}\Big]\,. \tag{7.17}$$

Proof. Summation of inequalities (7.5), (7.11), and (7.14), yields, after algebraic simplification, the variational inequality (7.15).

We now put variational inequality (7.15) into standard form which will be utilized in the subsequent sections. For additional background on variational inequalities and their applications, see the book by Nagurney [27]. In particular, we have that variational inequality (7.15) can be expressed as:

$$\langle F(X^*), X - X^* \rangle \geq 0, \quad \forall X \in k\,, \tag{7.18}$$

Where

$$X \equiv (Q^1, Q^2, \eta^1, \eta^2, \lambda, \rho_3) \text{ and } F(X) \equiv (F_{iljh}, F_{jkh\hat{l}}, \hat{F}_{iljh}, \hat{F}_{jkh\hat{l}}, F_j, F_{kh\hat{l}})$$

With indices: $i = 1, \ldots, I$; $l = 1, \ldots, L$; $j = 1, \ldots, J$; $h = 1, \ldots, H$; $\hat{l} = 1, \ldots, L$;, and the specific components of F given by the functional terms preceding the multiplication signs in (7.14), respectively. The term $\langle \cdot, \cdot \rangle$ denotes the inner product in N-dimensional Euclidean space.

We now describe how to recover the prices associated with the first two tiers of nodes in the global supply chain network. Such a pricing mechanism guarantees that the optimality conditions (7.5) and (7.11) as well as the equilibrium conditions (7.14) are satisfied individually through the solution of variational inequality (7.16).

Clearly, the components of the vector ρ_3^* are obtained directly from the solution of variational inequality ((7.16) as will be demonstrated explicitly through several numerical examples in Sect. 7.5). In order to recover the second tier prices associated with the retailers and the appreciation/exchange rates one can (after solving variational inequality (7.16) for the particular numerical problem) either (cf. (7.13) or (7.15)) set $\rho^{j*}_{2kh\hat{l}} \times e_h = \left[\rho^*_{3kh\hat{l}} - \hat{c}^j_{kh\hat{l}}(Q^{2*}, \eta^{2*})\right]$, for any $j, k, h, \hat{l}, m$ such that $q^{j*}_{kh\hat{l}} > 0$ or (cf. (7.11)) for any $q^{j*}_{kh\hat{l}} > 0$, set
$\rho^{j*}_{2kh\hat{l}} \times e_h = \left[\delta^j \frac{\partial r^j(Q^{1*}, Q^{2*}, \eta^{2*})}{\partial q^j_{kh\hat{l}}} + \frac{\partial c^j_{kh\hat{l}m}(q^{j*}_{kh\hat{l}}, \eta^{j*})}{\partial q^j_{kh\hat{l}}} + \lambda^*_j\right].$

Similarly, from (7.5) we can infer that the top tier prices can be recovered (once the variational inequality (7.16) is solved with particular data) thus:

for any i, l, j, h, m, such that
$q^{il*}_{jh} > 0$, set $\rho^{il*}_{1jh} \times e_h = \left[\frac{\partial f^{il}(q^{il*})}{\partial q^{il}_{jh}} + \alpha^{il} \frac{\partial r^{il}(Q^{1*}, \eta^{1*})}{\partial q^{il}_{jh}} + \frac{\partial c^{il}_{jh}(q^{il*}_{jh}, \eta^{il*})}{\partial q^{il}_{jh}}\right]$, or

equivalently, (cf. (7.11)), to $\left[\lambda_j^* - \delta^j \frac{\partial r^j(Q^{1*},Q^{2*},\eta^{2*})}{\partial q_{jh}^{il}} - \frac{\partial c_j(Q^{1*})}{\partial q_{jh}^{il}} - \frac{\partial \hat{c}_{jh}^{il}(q_{jh}^{il*},\eta^{j*})}{\partial q_{jh}^{il}}\right]$.

Under the above pricing mechanism, the optimality conditions (7.5) and (7.11) as well as the equilibrium conditions (7.15) also hold separately (as well as for each individual decision-maker) (see also, e. g., [8, 28]).

Existence of a solution to variational inequality (7.16) follows from the standard theory of variational inequalities, under the assumption that the functions are continuous, since the feasible set k is compact (cf. [27]). Also, according to the theory of variational inequalities, uniqueness of solution, in turn, is then guaranteed, provided that the function $F(X)$ that enters variational inequality (7.16) is strictly monotone on k.

Note that, if the equilibrium values of the flows (be they product or relationship levels) on links are identically equal to zero, then those links can effectively be removed from the network (in equilibrium). Moreover, the size of the equilibrium flows represents the "strength" of the respective links. In addition, the solution of the model reveals the true network structure in terms of the optimal relationships (and their sizes) as well as the optimal product transactions, and the associated prices.

7.2.9 Remark

We note that manufacturers as well as retailers may be faced with capacity constraints. Capacity limitations can be handled in the above model since the production cost functions, as well as the transaction cost functions and the handling cost functions can assume nonlinear forms (as is standard in the case of modeling capacities on roads in congested urban transportation networks (cf. [33])). Of course, one can also impose explicit capacity constraints and this would then just change the underlying feasible set(s) so that k would need to be redefined accordingly. However, the function $F(X)$ in variational inequality (7.18) would remain the same (see, e. g., [27]). Finally, since we consider a single homogeneous product the exchange rates e_h are assumed fixed (and relative to a base currency). Once can, of course, investigate numerous exchange rate and demand scenarios by altering the demand functions and the fixed exchange rates and then recomputing the new equilibrium product transaction, price, and relationship level equilibrium patterns.

7.3 The Supply Chain Network Efficiency and Vulnerability Measures

In this section, we propose the global supply chain network efficiency measure and the associated network component importance definition.

Definition 2 (The global supply chain network efficiency measure). The global supply chain network efficiency measure, ε, for a given network topology G, and demand price $\rho_{3kh\hat{l}}$, and available product from manufacturer il and retailer j, is defined as follows:

$$\varepsilon(G) = \frac{\sum_{k=1}^{K} \sum_{h=1}^{H} \sum_{\hat{l}=1}^{L} \frac{d_{kh\hat{l}}(\rho_3^*)}{\rho_{3kh\hat{l}}^*}}{K * H * L}, \tag{7.19}$$

where $K * H * L$ is the number of demand markets in the network, and $d_{kh\hat{l}}(\rho_3^*)$ and $\rho_{3kh\hat{l}}^*$ denote the equilibrium demand and the equilibrium price for demand market $kh\hat{l}$, respectively.

The global supply chain network efficiency measure, ε defined in (7.19) is actually the average demand to price ratio (cf. [32]). It measures the overall (economic) functionality of the global supply chain network. When the network topology G, the demand price functions, and the available product are given, a global supply chain network is considered to be more efficient if it can satisfy higher demand at lower prices.

By referring to the equilibrium conditions (7.13), we assume that if there is a positive transaction between a retailer with a demand market at the equilibrium, the price charged by the retailer plus the respective unit transaction costs is always positive. Hence, the prices paid by the demand market will always be positive and the above network efficiency measure is well-defined.

The importance of the network components is analyzed, in turn, by studying their impact on the network efficiency through their removal. The network efficiency of a global supply chain network can be expected to deteriorate when a critical network component is eliminated from the network. Such a component can include a link or a node or a subset of nodes and links depending on the network problem under study. Furthermore, the removal of a critical network component will cause more severe damage than that of a trivial one. Hence, the importance of a network component is defined as follows (cf. [32]):

Definition 3 (Importance of a global supply chain network component). The importance of network component g of global supply chain network G, $I(g)$, is measured by the relative global supply chain network efficiency drop after g is removed from the network:

$$I(g) = \frac{\varepsilon(G) - \varepsilon(G - g)}{\varepsilon(G)}, \tag{7.20}$$

where $\varepsilon(G - g)$ is the resulting global supply chain network efficiency after component g is removed.

It is worth pointing out that the above importance of the network components is well-defined even in a supply chain network with disconnected manufacturer/demand market pairs. In our supply chain network efficiency measure, the elimination of a transaction link is treated by removing that link from the network while the removal of a node is managed by removing the transaction links entering or exiting that node. The above procedure(s) to handle disconnected manufacturer/demand market pairs will be illustrated in the numerical examples in Sect. 7.4, when we compute the importance of the supply chain network components and their rankings.

Supply chain vulnerability arises from two sources, the risk within the supply chain and the external risk. The risk within the supply chain network is caused by sub-optimal interaction and co-operation between the entities along the chain. Such supply chain risks result from a lack of visibility, lack of 'ownership' and inaccurate forecasts. External risks are risks that arise from interactions between the supply chain and its environment. Such interactions include disruptions caused by strikes, terrorism and natural catastrophes. These risks impact the vulnerability of the supply chain. Thus, supply chain vulnerability can be defined as an exposure to serious disturbance, arising from risks within the supply chain as well as risks external to the supply chain.

Hence, vulnerability (ε_A) is a measure of the average decrease in efficiency of the network over the time horizon after attack and takes into account the entire history of the attack and the rapidity of the decline [3].

$$\varepsilon_A = \frac{\sum_{n=1}^{m} \varepsilon_n}{N} \tag{7.21}$$

Let $\varepsilon_1, \varepsilon_2, \ldots, \varepsilon_N$, be the efficiency the network after the elimination of 1, 2, up to N nodes. ε_A measures the average loss of efficiency during N attacks so that networks with a more rapid decline will have higher ε_A scores.

Consequently, supply chain risk management should aims at identifying the areas of potential risk and implementing appropriate actions to contain that risk. Therefore, we believe that the best way to mitigate supply chain risk and to reduce supply chain vulnerability as a whole is through coordinated and relationship approach amongst supply chain members.

7.4 Numerical Examples

In this section, we analyze the global supply chain networks efficiency and vulnerability. For each example, our network efficiency and vulnerability measures are computed and the importance and the rankings of nodes are also reported.

Example 1

The first set of numerical examples consisted of one country, two manufacturers, two currencies, two retailers, and two demand markets for the product. Hence, $L = 1$, $I = 2$, $H = 2$, $J = 2$, and $K = 2$, for this and the subsequent two numerical exam-

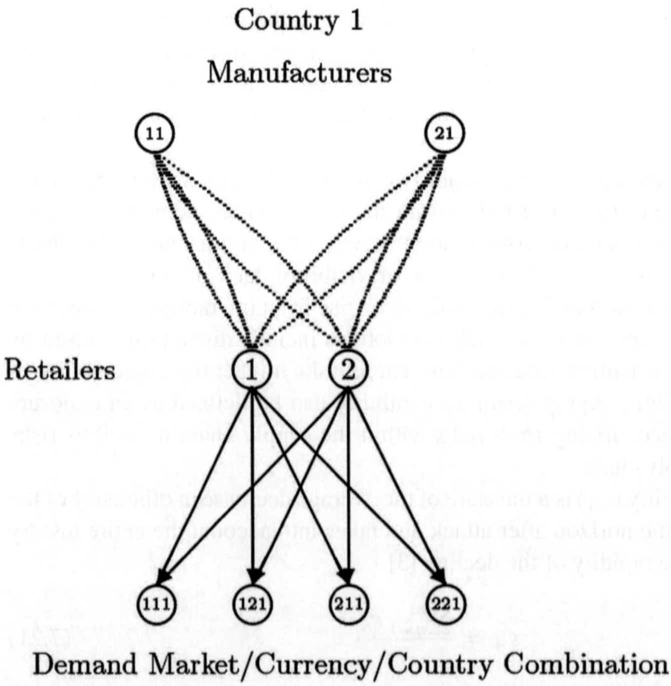

Fig. 7.2 Global supply chain network for Example 1

ples. The global supply chain network for the first example is depicted in Fig. 7.2. The examples below were solved using the Euler method (see [8, 28]).

The data for the first example were constructed for easy interpretation purposes (cf. Tables 7.2, 7.3, 7.5)

The transaction cost functions faced by the manufacturers associated with transacting with the retailers were given by:

$$c^{il}_{jh}(q^{il}_{jh}, \eta^{il}) = 0.5\left(q^{il}_{jh}\right)^2 + 3.5q^{il}_{jh} - \eta^{il}, \text{ for } i = 1,2; l = 1; j = 1,2; h = 1,2.$$

The production cost functions faced by the manufacturers were

$$f^{il}(q^{il}) = 0.5\left(\sum_{j=1}^{2}\sum_{h=1}^{2} q^{il}_{jh}\right)^2, \quad \text{for } i = 1,2; l = 1,2.$$

The handling costs of the retailers were given by:

$$c_j(Q^1) = .5\left(\sum_{j=1}^{2}\sum_{h=1}^{2} q^{il}_{jh}\right)^2, \quad \text{for } j = 1,2.$$

The transaction costs of the retailers associated with transacting with the manufacturers in the two countries were given by:

$$\hat{c}^{il}_{jh}\left(q^{il}_{jh}, \eta^{il}\right) = 1.5\left(q^{il}_{jh}\right)^2 + 3q^{il}_{jh}, \quad \text{for } i = 1,2; l = 1,2; j = 1,2; h = 1,2 .$$

The relationship cost functions were:

$$b^{il}(\eta^{il}) = 2\eta^{il}, \quad \text{for } i = 1,2,$$
$$b^{j}(\eta^{j}) = \eta^{j}, \quad \text{for } j = 1,2.$$

The demand functions at the demand markets were:

$$d_{111}(\rho_3) = -2\rho_{3111} - 1.5\rho_{3121} + 1000, \quad d_{121}(\rho_3) = -2\rho_{3121} - 1.5\rho_{3111} + 1000,$$
$$d_{211}(\rho_3) = -2\rho_{3211} - 1.5\rho_{3221} + 1000, \quad d_{221}(\rho_3) = -2\rho_{3221} - 1.5\rho_{3211} + 1000.$$

and the transaction costs between the retailers and the consumers at the demand markets (see (6.12)) were given by:

$$\hat{c}^{j}_{khl}(Q^2) = q^{j}_{khl} - \eta^{j} + 5, \quad \text{for } j = 1,2; k = 1,2; h = 1,2; l = 1 .$$

We assumed for this and the subsequent examples that the transaction costs as perceived by the retailers and associated with transacting with the demand markets were all zero, that is, $c^{j}_{khl}(q^{j}_{khl}) = 0$, for all j, k, h, l.

The Euler method converged and yielded the following equilibrium product shipment pattern:

$$q^{il*}_{jh} = 15.605, \quad \forall i, l, j, h , \quad q^{j*}_{khl} = 15.605 , \quad \forall j, k, h, l .$$

The vector γ^* had components: $\gamma^*_1 = \gamma^*_2 = 256.190$, and the computed demand prices at the demand markets were: $\rho^*_{3111} = \rho^*_{3121} = \rho^*_{3211} = \rho^*_{3221} = 276.797$.

Next, we analyze the importance of the network components by studying their impact on the network efficiency through their removal. Table 6 shows the results of this analysis.

Table 7.6 The importance of the supply chain decision-makers

Nodes	Importance Value	Ranking
Manufacturer (il)	0.3204	2
Retailer (j)	0.4854	1
Demand Market (khl)	−0.3086	3

Note that, given the cost structure and the demand price functions, since the retailer's nodes carry the largest amount of equilibrium product flow, they are ranked the most important. The negative importance values for demand markets are due to the fact that the existence of each demand market brings extra flows on the transaction links and nodes and, therefore, increases the marginal transaction cost. The removal of one demand market has two effects: first, the contribution to the network performance of the removed demand market becomes zero; second, the marginal transaction cost on links/nodes decreases, which decreases the equilibrium prices and increases the demand at the other demand markets. If the performance drop caused by the removal of the demand markets is overcompensated by the improvement of the demand-price ratio of the other demand markets, the removed demand market will have a negative importance value. It simply implies that the negative externality caused by the demand market has a larger impact than the performance drop due to its removal.

The vulnerability score for this supply chain network is $\varepsilon_A = 0.1657$.

Example 2

The first numerical example consisted of one country, two source agents, two currencies, two intermediaries, and two financial products. Hence, $L = 1$, $I = 2$, $H = 2$, $J = 2$, and $K = 2$. The network for the first example is depicted in Fig. 7.3. The data for this example is the same as in Example 1 except that in this example we allow the manufacturers to also transact directly with demand markets.

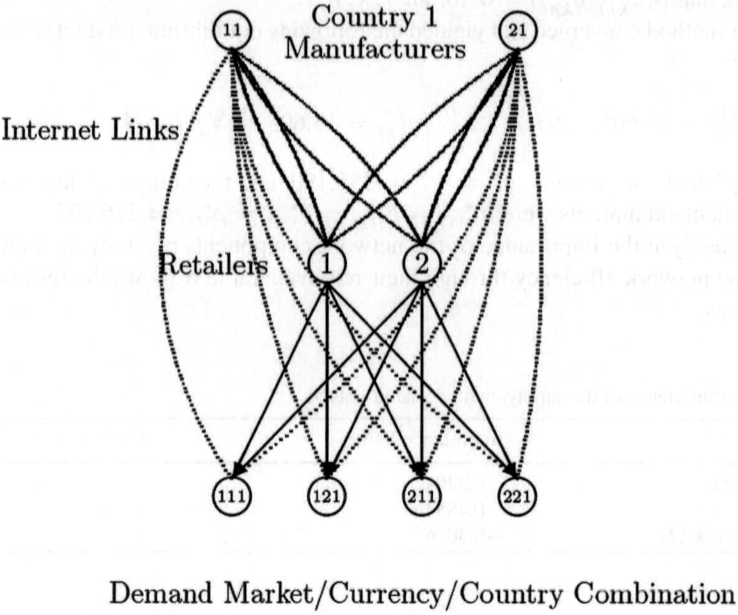

Fig. 7.3 Global supply chain network for Example 2

The transaction costs, in turn, associated with the transactions between manufacturers and the demand markets (from the perspective of the consumers) were given by:

$$\hat{c}^{il}_{kh\hat{l}}(q^{il}_{kh\hat{l}}) = 0.1q^{il}_{kh\hat{l}} + 1, \quad \forall i, l, \hat{l}, k, h\,.$$

The Euler method converged and yielded the following equilibrium supply chain flow pattern:

$$q^{il*}_{jh} = 2.017\,, \quad \forall i, l, j, h\,,$$
$$q^{il*}_{khl} = 26.82\,, \quad \forall i, l, k, h, l\,,$$
$$q^{j*}_{khl} = 2.017\,, \quad \forall j, k, h, l\,.$$

The vector γ^* had components: $\gamma_1^* = \gamma_2^* = 268.58$, and the computed demand prices at the demand markets were: $\rho^*_{3111} = \rho^*_{3121} = \rho^*_{3211} = \rho^*_{3221} = 269.236$. The importance of the nodes of this network and their ranking are reported in Table 7.7.

For this network the most important nodes are the manufacturers nodes since they transact and have relationships with retailers and demand markets. The vulnerability score for this supply chain network is $\varepsilon_A = 0.019$.

Table 7.7 The importance of the supply chain decision-makers

Nodes	Importance Value	Ranking
Manufacturer (il)	0.486	1
Retailer (j)	−0.03	2
Demand Market (khl)	−0.19	3

Example 3

In this numerical example, the global supply chain network was as given in Fig. 7.4. This example consisted of two countries with two manufacturers in each country; two currencies, two retailers, and two demand markets. Hence, $L = 2$, $I = 2$, $H = 2$, $J = 2$, and $K = 2$.

The data for this example is the replication of the data for Example 1, from 1 country to two countries. Therefore, the demand functions at the demand markets were:

$$d_{111}(\rho_3) = -2\rho_{3111} - 1.5\rho_{3121} + 1000\,, \quad d_{121}(\rho_3) = -2\rho_{3121} - 1.5\rho_{3111} + 1000\,,$$
$$d_{211}(\rho_3) = -2\rho_{3211} - 1.5\rho_{3221} + 1000\,, \quad d_{221}(\rho_3) = -2\rho_{3221} - 1.5\rho_{3211} + 1000\,,$$
$$d_{112}(\rho_3) = -2\rho_{3112} - 1.5\rho_{3122} + 1000\,, \quad d_{122}(\rho_3) = -2\rho_{3122} - 1.5\rho_{3112} + 1000\,,$$
$$d_{212}(\rho_3) = -2\rho_{3212} - 1.5\rho_{3222} + 1000\,, \quad d_{222}(\rho_3) = -2\rho_{3222} - 1.5\rho_{3212} + 1000\,.$$

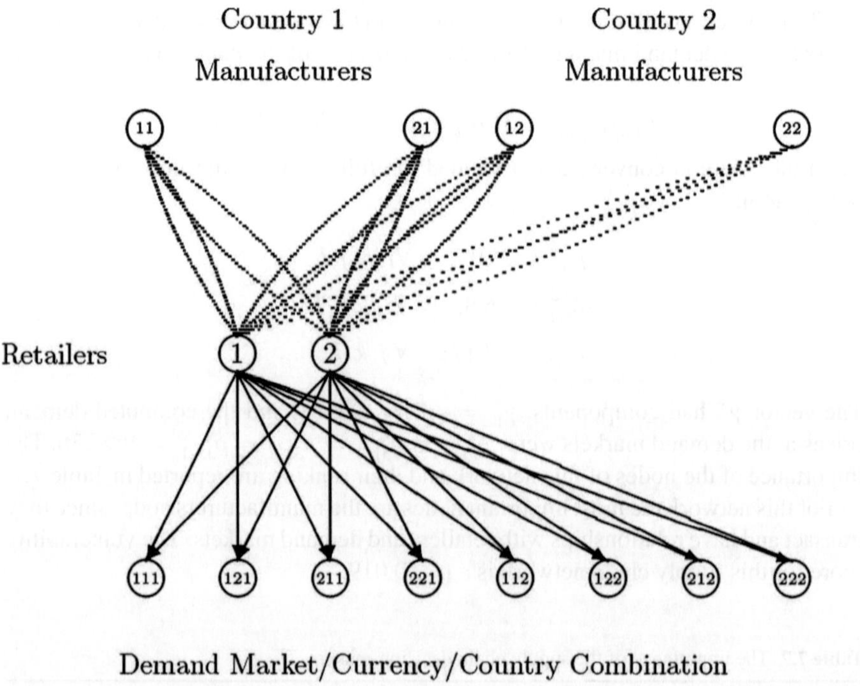

Fig. 7.4 Global supply chain network for Example 3

The Euler method converged (in 318 iterations) and yielded the following equilibrium product shipment pattern:

$$q_{jh}^{il*} = 16.305\,, \quad \forall i,l,j,h\,, \quad q_{khl}^{j*} = 16.305\,, \quad \forall j,k,h,l\,.$$

The vector γ^* had components: $\gamma_1^* = \gamma_2^* = 266.882$ and the computed demand prices at the demand markets were: $\rho_{3111}^* = \rho_{3121}^* = \rho_{3211}^* = \rho_{3221}^* = \rho_{3212}^* = \rho_{3222}^* = 277.379$.

The importance of the nodes of this network and their ranking are reported in Table 7.8. The vulnerability score for this supply chain network is $\varepsilon_A = 0.060$. This network is less vulnerable than the one from Example 1, since the retailers can get their products from many more manufacturers and sell them to many more demand markets as compare to the retailers in Example 1.

Table 7.8 The importance of the supply chain decision-makers

Nodes	Importance Value	Ranking
Manufacturer (il)	0.18	2
Retailer (j)	0.42	1
Demand Market (khl)	−0.307	3

We note that supply chain in Example 1 is the most vulnerable, follow by the network in Example 3. A vulnerable supply chain would have a high vulnerability score. Network in Example 2 is the least vulnerable since manufacturers have more options to sell their products, to retailers and demand market. The results in this chapter can be used to assess which nodes and links in supply chain networks are the most vulnerable in the sense that their removal will impact the performance of the network in the most significant way. Moreover, for each member of the supply chain network the highest is it relationship or connectivity the lowest is its vulnerability. Therefore, supply chain risk management should be based on clear performance requirements and lines of communication or relationship between all members of the chain. It is important to note that the vulnerability of the supply chain depends on the design and structure of the network and on the type of relationship between its members.

7.5 Managerial Implications

As managers determine the appropriate practices to manage the supply chain efficiency and vulnerability, the supply chain design and structure should take in consideration risk issues. When considering risk issues it is important to the firm to create a portfolio of relationships. Strong supply chain relationships enable firms to react to changes in the market, improve forecasting and create supply chain visibility, which lead to improve profit margins [11]. The benefits are reduction of production, transportation and administrative costs. Moreover, multiple relationships can help companies deal with the negative consequences related to dependence on supply chain partners. Thus, supply chain vulnerability can be minimized as an exposure to serious disturbance is reduced. Companies will be able to deal with disturbance arising from risks within the supply chain as well as risks external to the supply chain when there is an optimal interaction and co-operation between the entities along the chain.

The benefits of investing in social relationship are not only economical but also technical and social. On the technical development the greatest benefit is the possibility of sharing the resources of suppliers and shortening the lead-times. Spekman and Davis [35] found that supply chain networks that exhibit collaborative behaviors tend to be more responsive and that supply chain-wide costs are, hence, reduced. These results are also supported by our economic model where we demonstrated that a higher level of relationship lowers transaction costs and risk and uncertainty. As a result, supply chain prices are reduced and the overall transactions increases.

Risk management in supply chain should be based on clear performance requirements, optimal level of investment in supply chain relationships, and on process alignment and cooperation within and between the entities in the supply chain. Therefore, the structure and design of the supply chain will determine how vulnerable it will be.

7.6 Conclusions

In this chapter, we develop a framework for the analysis of the optimal levels of social relationship in a global supply chain network consisting of manufacturers, retailers, and consumers. We propose a network performance measure for the evaluation of supply chain networks efficiency and vulnerability. The measure captures risk, transaction cost, price, transaction flow, revenue, and demand information in the context of the decision-makers behavior. Manufacturers and retailers are multicriteria decision-makers who decide about their production and transaction quantities as well as the amount of social relationship they want to pursue in order to maximize net return and minimize risk.

We construct the finite-dimensional variational inequality governing the equilibrium of the competitive global supply chain network. The model allows us to investigate the interplay of the heterogeneous decision-makers in the global supply chain and to compute the resultant equilibrium pattern of product outputs, transactions, product prices, and levels of social relationship. A computational procedure that exploits the network structure of the problem is applied to several numerical examples.

Results of our numerical examples highlight the importance of considering the impact of relationship levels in a global supply chain context. Furthermore, they stress the importance of a network perspective, the importance of each decision makers in the network and the overall efficiency and vulnerability of the global supply chain network . These examples, although stylized, have been presented to show both the model and the computational procedure. Obviously, different input data and dimensions of the problems solved will affect the equilibrium product transaction, levels of social relationship, price patterns, and the supply chain vulnerability.

Future research will extend this framework to include other criteria and the introduction of dynamics. Future research may also focus on the study of type of relationship investments required in each stages of the supply chain network. It is important to know to what extent do the relationships between elements in supply chain differ depending on the stage of the alliance and on aspects of the costs and benefits of the relationship. How do the combinations of these elements play in each time period and stage of the supply chain? These questions await empirical study. We feel we have only scratched the surface and look forward to future studies that will help researchers better conceptualize and theorize specific aspect of supply chain relationship, vulnerability and risk management.

References

1. Baker WE, Faulkner RR (2004) Social networks and loss of capital. Soc Networks 26:91–111
2. Bazaraa MS, Sherali HD, Shetty CM (1993) Nonlinear programming: Theory and algorithms. John Wiley & Sons, New York

3. Bienenstock EJ, Bonacich P (2003) Balancing efficiency and vulnerability in social networks. In: Breiger R , Carley K, Pattison P (eds) Dynamic social network modeling and analysis: Workshop summary and papers. The National Academies Press, Washington, DC, 253–264
4. Cannon JP, Perreault Jr WD (1999) Buyer-seller relationships in business markets. J Marketing Res 36:439–460
5. Christopher M, Jüttner U (2000) Developing strategic partnerships in the supply chain: A practitioner perspective. Eur JPurchasing Supply Manag 6: 117–127
6. Christopher M, Peck H (2004) Building the resilient supply chain. Int J Logist Manag 15(2):1–13
7. Crosby LA, Stephens N (1987) Effects of relationship marketing on satisfaction, retention, and prices in the life insurance industry. J Marketing Res 24:404–411
8. Cruz JM, Nagurney A, Wakolbinger T (2006) Financial engineering of the integration of global supply chain networks and social networks with risk management. Nav Res Logist 53:674–696
9. Dyer J (2000) Collaborative advantage . Oxford University Press, Oxford, England
10. Engardio P, Shari M, Weintraub A, Arnst C (2003) Deadly virus, BusinessWeek Online; see: http://businessweek.com
11. Flint DJ (2004) Strategic marketing in global supply chains: four challenges. Ind Market Manag 33:45–50
12. Fisher ML (1997) What is the right supply chain for your product? Harvard Bus Rev 75(2):105–116
13. Gabarino E, Johnson MS (1999) The different roles of satisfaction, trust, and commitment in customer relationships. J Marketing 63:70–87
14. Gadde L, Snehota I (2000) Making the most of supplier relationships. Ind Market Manag 29:305–316
15. Golicic SL, Foggin JH, Mentzer JT (2003) Relationship magnitude and its role in interorganizational relationship structure. J Bus Logist 24:57–75
16. Granovetter M (1985) Economic action and social structure: The problem of embeddedness. Am JSociol 91:481–510
17. Harland C, Brenchley R, Walker H (2003) Risk in supply networks. J Purchasing and Supply Manag 9(2):51–62
18. Hendricks KB, Singhal VR (2005) An empirical analysis of the effects of supply chain disruptions on long-run stock price performance and equity risk of the firm. Prod Oper Manag 14(1):35–52
19. Hult GT M, Ketchen Jr DJ, Slater SF (2004) Information processing, knowledge development, and strategic supply chain performance. Acad Manag J 47(2):241–253
20. Husted BW (1994) Transaction costs, norms, and social networks. Bus Soc 33:303–324
21. Jüttner U, Peck H, Christopher M (2003) Supply chain risk management: Outlining and agenda for future research. Int J of Logist: Res Applications 6:197–210
22. Johnson ME (2001) Learning from toys: Lessons in managing supply chain risk from toy industry. Calif Manage Rev 43:106–130
23. Keeney RL, Raiffa H (1993) Decisions with multiple objectives: Preferences and value trade-offs. Cambridge University Press, Cambridge, England
24. Kleindorfer PR, Saad GH (2005) Managing disruption risks in supply chains. Prod Oper Manag 14(1):53–68
25. Krause DR, Handfield RB, Tyler BB (2007) The relationships between supplier development, commitment, social capital accumulation and performance improvement. J Oper Manag 25(2):528–545
26. Martha J, Subbakrishna S (2002) Targeting a just-in-case supply chain for the inevitable next disaster. Supply Chain Manag Rev 6(5):18–23
27. Nagurney A (1999) Network economics: A variational inequality approach (Second and revised edition). Kluwer Academic Publishers, Dordrecht, The Netherlands

28. Nagurney A, Cruz J, Matsypura D (2003) Dynamics of global supply chain supernetworks. Math Comput Model 37:963–983
29. Nash JF (1950) Equilibrium points in n-person games. P Natl Acad Sci 36:48–49
30. Nash JF (1951) Noncooperative games. Ann Math 54:286–298
31. Norrman A, Jansson U (2004) Ericssons proactive supply chain risk management approach after a serious sub-supplier accident. Int J Phys Distrib Logist Manag 34:434–456
32. Qiang Q, Nagurney A (2008) A unified network performance measure with importance identification and the ranking of network components Optim Lett 2:127–142
33. Sheffi Y (1985) Urban transportation networks – Equilibrium analysis with mathematical programming methods. Prentice-Hall, Englewood Cliffs, New Jersey
34. Sheffi Y (2001) Supply chain management under the threat of international terrorism. Int J Logist Manag 12:1–11
35. Spekman RE, Davis EW (2004) Risk business: Expanding the discussion on risk and the extended enterprise. Int J Phys Distrib Logist Manag 34:414–433
36. Tang CS (2006) Robust strategies for mitigating supply chain disruptions. Int J Logist: Res Applications 9(1):33–45
37. Uzzi B (1997) Social structure and competition in interfirm networks: The paradox of embeddedness. Admin Sci Quart 42:35–67
38. Wakolbinger T, Nagurney A (2004) Dynamic supernetworks for the integration of social networks and supply chains with electronic commerce: Modeling and analysis of buyer-seller relationships with computations. Netnomics 6:153–185
39. Zsidisin GA, Panelli A , Upton R (2000) Purchasing organization involvements in risk assessment, contingency plans, and risk management: an exploratory study. Supply Chain Manag 5(4):187–197

Chapter 8
A Stochastic Model for Supply Chain Risk Management Using Conditional Value at Risk

Mark Goh and Fanwen Meng

Abstract In this chapter, we establish a stochastic programming formulation for supply chain risk management using conditional value at risk. In particular, we investigate two problems on logistics under conditions of uncertainty. The sample average approximation method is introduced for solving the underlying stochastic model. Preliminary numerical results are provided.

8.1 Introduction

For supply chains that comprise hundreds of companies and several tiers, there are numerous risks to consider. Generally, these risks can be classified into two types: risks arising from within the supply chain (operational risk) and risks external to the chain (disruption risk). The attributes of operational risks are due to the interactions between firms across the supply chain, such as supply risk, demand risk, and trade credit. Disruption risks arise from the interactions between the supply chain and its environment, such as terrorism, or natural disasters such as the severe acute respiratory syndrome (SARS). Therefore, supply chain risk management can be defined as the identification and management of operational risks and disruption risks through a coordinated approach amongst supply chain members to reduce supply chain vulnerability. For recent related work, see [5] and the references therein. A survey on studies in this regard can be found in Tang [25].

In stochastic optimization, Value at Risk (VaR) is a widely used risk measure for quantifying the downside risk. A drawback of this approach is that it often destroys the convexity of the model, which makes the resulting optimization problem hard to solve numerically. An extensive recent study on a related measure of risk, Conditional Value at Risk (CVaR), popularized by Rockafellar and Uryasev in [18, 19], as the tightest convex approximation of VaR, has shown that CVaR is a coherent risk measure that enjoys convexity and subadditivity. Therefore, CVaR

T. Wu, J. Blackhurst (eds.), *Managing Supply Chain Risk and Vulnerability*

has advantages over VaR at least in numerical computation. Due to this, in this chapter, we use CVaR as a risk measure in the framework of supply chain risk management. For a detailed discussion on VaR and CVaR, the reader is referred to [1, 2, 4, 7, 8, 9, 10, 14, 18, 19, 24, 27, 28].

Suppose $f : R^n \times R^m \to R$ is a loss function which depends on the control vector $x \in X \subseteq R^n$ and the random vector $y \in R^m$. Here X denotes the set of decision constraints and $y : \Omega \to \Xi \subseteq R^m$ is a random vector defined on probability space (Ω, F, P), which is independent of x. Here x is a decision vector and y serves as a vector representing uncertainties, e. g., the risks involved in supply chains. Then, for each $x \in X$, the loss $z = f(x, y)$ is a random variable with a distribution in R induced by that of y. Recall that for a given confidence level $\alpha \in (0, 1)$, the VaR of the loss z associated with a decision x is defined as

$$\mathrm{VaR}_\alpha(x) = \min\{u | P\{y | f(x, y) \le u\} \ge \alpha\}\ , \tag{8.1}$$

which is the minimum value of the loss at a point such that the probability of the loss is less than or equals to α. CVaR is defined as the expectation of the loss $f(x, y)$ in the conditional distribution of its upper α-tail. A more *operationally convenient* definition [18, 19] of CVaR, as the optimal value function of an unconstrained minimization problem, is as follows.

$$\mathrm{CVaR}_\alpha(x) = \min\{\eta_\alpha(x, u) | u \in R\}\ , \tag{8.2}$$

where

$$\eta_\alpha(x, u) := u + \frac{1}{1 - \alpha} E\,[f(x, y) - u]^+\ , \tag{8.3}$$

where the superscript plus sign denotes the nonnegative part of a scalar, i. e., the *plus function* $[t]^+ := \max\{0, t\}$. Note that $\mathrm{VaR}_\alpha(x)$ is the lower (or left) endpoint of the closed bounded interval $\arg\min_u \eta_\alpha(x, u)$ and

$$\mathrm{CVaR}_\alpha = \eta_\alpha(x,\ \mathrm{VaR}_\alpha(x))\ . \tag{8.4}$$

Further, Rockafellar and Uryasev [18, 19] showed that the CVaR minimization problem

$$\min\{\mathrm{CVaR}_\alpha(x) | x \in X\} \tag{8.5}$$

is equivalent to

$$\min\{\eta_\alpha(x, u) | (x, u) \in X \times R\}\ , \tag{8.6}$$

in the sense that these two problems achieve the same minimum value.

In general, the loss function f is assumed to be linear in x. To deal with the plus function involved in the CVaR minimization problem, some auxiliary variables are often introduced. Then, algorithms based on linear programming can be employed for solving the underlying CVaR minimization problem [1, 4, 14, 28]. However, introducing additional variables can increase the size of the problem greatly as the sample size of the random variable y increases. This might result in a big challenge in computation. In this chapter, we assume that f is a general convex loss function.

In other words, f can be a nonlinear function, such as a quadratic function, or a piecewise linear/quadratic function.

In Sect. 8.2, we establish a general stochastic model for supply chain risk management using CVaR. A sampling approach called Sample Average Approximation (SAA) is introduced to solve the stochastic model in Sect. 8.3. SAA methods have been extensively investigated in stochastic optimization [3, 6, 23]. Recently, the SAA method was applied for supply network design problems under uncertainty [21] and stochastic mathematical programs with equilibrium constraints [12, 23, 29]. Based on the general stochastic model and the SAA method discussed in Sects. 8.2 and 8.3, we investigate an example of a supply chain consisting of two suppliers, one contract manufacturer, and two distributors as an illustration in Sect. 8.4. We investigate the risk management of the wine supply chain where the uncertain demand follows a discrete distribution in Sect. 8.5. Section 8.6 concludes.

8.2 A Stochastic Model with Conditional Value at Risk

The supply chain under consideration consists of one sole decision maker, a manufacturer, and a number of tiers of finitely many players in the chain which are closely connected to the manufacturer. For example, in the two-tier case, the players of the upstream tier could be suppliers who ship raw materials to the manufacturer, while the players from the downstream tier are the distributors of the manufactured products. At each pair, there are uncertainties and risks involved, ranging from supply of raw material, production quality, and demand.

We formulate a stochastic model of supply chain risk management using the risk measure of CVaR. We suppose that the associated risks, denoted by a random vector $y \in R^m$, of the underlying supply chain, follow a certain distribution. Let $x \in R^n$ denote the decision variable, X denote the basic domain of decision variables, such as, the nonnegative orthant R^n_+ or some bounded region in R^n, and $\varphi(x, y)$ the profit function of the decision maker. Then, the corresponding loss function $f(x, y) := -\varphi(x, y)$.

The decision maker aims to both maximize his expected profit and minimize his expected loss, subject to some constraints. In general, the decision maker might set a prior loss tolerance, β, concerning the value of risk. We then set the following constraint of the model:

$$\text{CVaR}_\alpha(x) \leq \beta\,, \tag{8.7}$$

where α is the confidence level given by the decision maker.

Further, let $h_l(x, y) : R^n \times \Omega \to R$, $l = 1, \ldots, k$, denote the constraints, representing the capacities, the product flows, and other activity constraints with uncertain demand and other risks involved in the chain. In this chapter, we are interested in the mathematical expectations of such underlying constraints, i. e., $\mathrm{E}\,[h_l(x, y)]$, although the decision maker might have an overly optimistic or pessimistic attitude as facing some extreme value distributions of uncertainties y. We assume the expectations of such constraints are satisfied by the following requirements:

$$\mathrm{E}\,[h_l(x,y)] = 0\,, \quad l = 1, 2, \ldots, j\,, \tag{8.8}$$

$$\mathrm{E}\,[h_l(x,y)] \le 0\,, \quad j = l+1,\ \ldots, k\,. \tag{8.9}$$

We now formulate the objective function of the model. The motivation and analysis is as follows. It is known that the attitude of the decision makers has a critical role in the decision process. An optimistic or pessimistic attitude can lead to an extreme solution of the system. However, finding such an optimal solution is costly and might even lead to additional cost to the decision makers under certain situations such as incomplete market information. Due to this, we ignore all analysis related to extreme value distributions. Instead, we consider the expected objective in the analysis. For an extensive study and application of this approach, we refer to recent publications in stochastic programming, such as [12, 22, 29].

In general, two criteria are considered for the decision maker, that is, (i) maximizing the profit; (ii) minimizing the loss. To characterize these two criteria in the objective function simultaneously, we introduce a simple *risk averse rate* $\lambda \in [0,1]$ and define the objective as a convex combination of the two criteria, namely, $(1-\lambda)\mathrm{E}\,[f(x,y)] + \lambda \mathrm{CVaR}_\alpha(x)$.

Thus, the stochastic model can be formulated as:

$$\begin{aligned} &\min (1-\lambda)\mathrm{E}\,[f(x,y)] + \lambda \mathrm{CVaR}_\alpha(x) \\ s.t.\ \ &\mathrm{E}\,[h_l(x,y)] = 0\,, \quad l = 1, 2, \ldots, j\,, \\ &\mathrm{E}\,[h_l(x,y)] \le 0\,, \quad l = j+1, \ldots, k\,, \\ &\mathrm{CVaR}_\alpha(x) \le \beta\,, \quad x \in X\,. \end{aligned} \tag{8.10}$$

Note that, by (8.2), $\mathrm{CVaR}_\alpha(x)$ is the optimal value function of an unconstrained minimization problem, making the resulting problem (8.10) hard or even impossible to solve directly. To overcome this obstacle in computation, we establish an equivalent formulation of problem (8.10). Before stating Proposition 1, we need the following lemma, which is taken from ([18], Theorem 10).

Lemma 1. *Suppose that* $\mathrm{E}\,[|f(x,y)|] < \infty$ *for any* $x \in X$. *Then, for* $x \in X$, $\eta_\alpha(x,u)$, *as a function of* $u \in R$, *is finite and convex.*

Proposition 1. Suppose that $\mathrm{E}\,[|f(x,y)|] < \infty$ for any $x \in X$. The minimization problem (8.10) is equivalent to

$$\begin{aligned} &\min (1-\lambda)\mathrm{E}\,[f(x,y)] + \lambda \left(u + \frac{1}{1-\alpha}\mathrm{E}\,[f(x,y)-u]^+ \right) \\ s.t.\ \ &\mathrm{E}\,[h_l(x,y)] \le 0\,, \quad l = j+1, \ldots, k\,, \\ &u + \frac{1}{1-\alpha}\mathrm{E}\,[f(x,y)-u]^+ \le \beta, \quad (x,u) \in X \times R\,, \end{aligned} \tag{8.11}$$

in the sense that their objectives achieve the same minimum value. Moreover, if the CVaR *constraint in (8.10) is active,* (x^*, u^*) *achieves the minimum of (8.11) if and only if* x^* *achieves the minimum of (8.10) with*

$$u^* \in \arg\min_{u \in R} \left\{ u + \frac{1}{1-\alpha} \mathrm{E}\left[f(x^*, y) - u\right]^+ \right\} .$$

Proof. To ease notation, set $\hat{X} = X \cap \{x \in R^n | \mathrm{E}[h_l(x, y)] = 0, l = 1, 2, \ldots, j, \mathrm{E}[h_l(x, y)] \le 0, \quad l = j+1, \ldots, k\}$. *In the following, we consider two cases: (i)* $\lambda = 0$*; (ii)* $0 < \lambda \le 1$.

For case (i), it is evident that (8.10) is reduced to minimize $\mathrm{E}[f(x, y)]$ *over* $\hat{X}$ *subject to* $\mathrm{CVaR}_\alpha(x) \le \beta$*, and (8.11) becomes minimizing* $\mathrm{E}[f(x, y)]$ *over* $\hat{X} \times R$ *subject to* $\eta_\alpha(x, u) \le \beta$*. This is exactly the same situation considered in [18]. The desired result follows directly by ([18], Theorem 16).*

We now consider case (ii). If (x^*, u^*) *solves (8.11), then* $\eta_\alpha(x^*, u^*) \le \beta$*, and*

$$\begin{aligned} &(1-\lambda)\mathrm{E}\left[f(x^*, y)\right] + \lambda\eta_\alpha(x^*, u^*) = \\ &\min_{u \in R} \left\{(1-\lambda)\mathrm{E}\left[f(x^*, y)\right] + \lambda\eta_\alpha(x^*, u) | \eta_\alpha(x^*, u) \le \beta\right\} . \end{aligned}$$

This implies that for any

$$\begin{aligned} &u \in \hat{U} := \{u \in R | \eta_\alpha(x^*, u) \le \beta\} , \\ &(1-\lambda)\mathrm{E}\left[f(x^*, y)\right] + \lambda\eta_\alpha(x^*, u^*) \le (1-\lambda)\mathrm{E}\left[f(x^*, y)\right] + \lambda\eta_\alpha(x^*, u) . \end{aligned}$$

With $0 < \lambda \le 1$*, it then follows that* $\eta_\alpha(x^*, u^*) \le \eta_\alpha(x^*, u)$ *for any* $u \in \hat{U}$*. Thereby,* $\eta_\alpha(x^*, u^*) = \min\left\{\eta_\alpha(x^*, u) | u \in \hat{U}\right\}$*. In addition, by Lemma 1,* $\eta_\alpha(x^*, u)$ *is finite and convex in* u*. Thus,* $u \in \hat{U}$*. Then,*

$$\eta_\alpha(x^*, u^*) = \min_{u \in \hat{U}} \eta_\alpha(x^*, u) = \min_{u \in R} \eta_\alpha(x^*, u) = \mathrm{CVaR}_\alpha(x^*) ,$$

which implies $u^* \in \arg\min_{u \in IR} \eta_\alpha(x^*, u)$. Next, we show x^* solves (8.10). According to the equivalent expression (8.2) of CVaR, for any $x \in \hat{X}$, $\mathrm{CVaR}_\alpha(x) \le \beta$ if and only if there exists $u(x)$ such that $\eta_\alpha(x, u(x)) = \mathrm{CVaR}_\alpha(x) \le \beta$. Thereby, $(x, u(x))$ is a feasible point to problem (8.11). Further, note that x^* is feasible to (8.10) since $\mathrm{CVaR}_\alpha(x^*) = \eta_\alpha(x^*, u^*) \le \beta$. Also, (x^*, u^*) is an optimal solution of (8.11). Then, for any $x \in \hat{X}$ satisfying $\mathrm{CVaR}_\alpha(x) \le \beta$ we have

$$\begin{aligned} &(1-\lambda)\mathrm{E}\left[f(x^*, y)\right] + \lambda\mathrm{CVaR}_\alpha(x^*) = (1-\lambda)\mathrm{E}\left[f(x^*, y)\right] + \lambda\eta_\alpha(x^*, u^*) \\ &\le (1-\lambda)\mathrm{E}[f(x, y)] + \lambda\eta_\alpha(x, x(u)) = (1-\lambda)\mathrm{E}[f(x, y)] + \lambda\mathrm{CVaR}_\alpha(x) . \end{aligned}$$

Thereby, x^*solves (8.10).

On the other hand, suppose that x^* solves (8.10). Let $u^* \in \arg\min_{u \in R} \eta_\alpha(x^*, u)$. Then, $\mathrm{CVaR}_\alpha(x^*) = \eta_\alpha(x^*, u^*) \le \beta$. This implies that (x^*, u^*) is feasible to (8.11). Note that, for $x \in \hat{X}$, there exists $u(x) \in R$, such that $\mathrm{CVaR}_\alpha(x) = \eta_\alpha(x, u(x))$. Then, for any (x, u) feasible to (8.11), i. e., $(x, u) \in \hat{X} \times R$ satisfying $\eta_\alpha(x, u) \leq \beta$, it follows that

$$\begin{aligned}\mathrm{CVaR}_\alpha(x) &= \eta_\alpha(x, u(x)) = \min\{\eta_\alpha(x, u) | u \in R\}\\ &\leq \min\{\eta_\alpha(x, u) | u \in R, \eta_\alpha(x, u) \leq \beta\} \leq \beta\,.\end{aligned}$$

Thus, x is feasible to (8.10). As x^* is a solution of (8.10), then for any $(x, u) \in \hat{X} \times R$ feasible to (8.11),

$$\begin{aligned}&(1-\lambda)\mathrm{E}\left[f(x^*, y)\right] + \lambda \mathrm{CVaR}_\alpha(x^*) = (1-\lambda)\mathrm{E}\left[f(x^*, y)\right] + \lambda \eta_\alpha(x^*, u^*)\\ &\quad \leq (1-\lambda)\mathrm{E}\left[f(x, y)\right] + \lambda \mathrm{CVaR}_\alpha(x) = (1-\lambda)\mathrm{E}\left[f(x, y)\right] + \lambda \eta_\alpha(x, u(x))\\ &\quad = (1-\lambda)\mathrm{E}\left[f(x, y)\right] + \lambda \min\{\eta_\alpha(x, u) | u \in R\}\\ &\quad \leq (1-\lambda)\mathrm{E}\left[f(x, y)\right] + \lambda \min\{\eta_\alpha(x, u) | u \in R, \eta_\alpha(x, u) \leq \beta\}\\ &\quad \leq (1-\lambda)\mathrm{E}\left[f(x, y)\right] + \lambda \eta_\alpha(x, u)\,.\end{aligned}$$

Thus, (x^*, u^*) solves (8.11). This completes the proof.

In this chapter, for convenience, we assume that all the underlying expected functions in (8.11) are finite and continuous. (8.11) is the usual (one-stage) stochastic programming problem. Hence, it might be possible to apply some well developed optimization approaches in stochastic programming to solve this problem. However, a challenge is on how to deal with the mathematical expectations. In Sect. 8.3, we will introduce a sampling approach method for solving (8.11). If λ is chosen to be 1, the decision maker wishes to solely minimize his loss or risk. This is viewed as a conservative strategy. In this case, the original problem (8.10) is reduced to

$$\begin{aligned}&\min \mathrm{CVaR}_\alpha(x)\\ &s.t. \quad \mathrm{E}\left[h_l(x, y)\right] = 0\,, \quad l = 1, 2, \ldots, j\,,\\ &\mathrm{E}\left[h_l(x, y)\right] \leq 0\,, \quad l = j+1, \ldots, k\,,\\ &\mathrm{CVaR}_\alpha(x) \leq \beta\,, \quad x \in X\,.\end{aligned} \tag{8.12}$$

Likewise, if $\lambda = 0$, the decision maker treats the expected profit more important than the potential loss. In practice, the decision maker can choose a value of $\lambda \in [0, 1]$ based on his own preference for risk.

8.3 Sample Average Approximation Program

We now discuss numerical methods for solving (8.11). We introduce a Monte Carlo simulation method, the sample average approximation method. SAA methods have been extensively investigated in stochastic optimization in the past few years [22]. More recently, the SAA method has been applied to supply network design problems under uncertainty [21] and stochastic mathematical programs with equilibrium constraints [12, 29].

Let $\{y^1, y^2, \ldots, y^N\}$ be an independent identically distributed (i.i.d) sample of y. Define

$$\bar{f}_N(x) := \frac{1}{N}\sum_{i=1}^{N} f(x, y^i),$$

$$\hat{f}_N(x,u) := \frac{1}{N}\sum_{i=1}^{N} \left[f(x, y^i) - u\right]^+,$$

$$\bar{h}_{lN}(x) := \frac{1}{N}\sum_{i=1}^{N} h_l(x, y^i), \quad l = 1, \ldots, k.$$

The SAA program of problem (8.11) is given as follows.

$$\begin{aligned} &\min (1-\lambda)\bar{f}_N(x) + \lambda\left(u + \frac{1}{1-\alpha}\bar{f}_N(x,u)\right) \\ &\quad s.t. \quad \bar{h}_{lN}(x,y) = 0, \quad l = 1, 2, \ldots, j, \\ &\qquad\qquad \bar{h}_{lN}(x,y) \le 0, \quad l = j+1, \ldots, k, \\ &u + \frac{1}{1-\alpha}\hat{f}_N(x,u) \le \beta, \quad x \in X, u \in R. \end{aligned} \tag{8.13}$$

Note that the above SAA program is a deterministic nonlinear programming problem. Thus, we can employ some well developed methods in nonlinear programming to solve (8.13). According to ([22], Proposition 7), we derive the following results.

Proposition 2. *Let Z be a nonempty compact subset of $R^n \times R$. Suppose that (i) $f(\cdot, y)$ and $h_l(\cdot, y)$ are continuous on Z for a.e. $y \in \Omega$; (ii) $f(x, \cdot)$, $[f(x, \cdot) - u]^+$, $h_l(x, \cdot)$, are dominated by an integrable function, for any $(x, u) \in Z$ and $l = 1, \ldots, k$. Then, $\bar{f}_N(x)$, $\hat{f}_N(x, u)$, and $\bar{h}_{lN}(x)$ converge to $\mathrm{E}\left[f(x, y)\right]$, $\mathrm{E}[f(x,y) - u]^+$, and $\mathrm{E}\left[h_l(x, y)\right]$, $l = 1, \ldots, k$, w.p.1 uniformly on Z, respectively.*

To ease notation, we denote by v^*, S, $\bar{v}_N$, and $\bar{S}_N$ the optimal value, the set of optimal solutions of (8.11), the optimal value, and the set of optimal solutions of (8.13), respectively. We derive the following proposition concerning the convergence of the optimal value of the SAA program.

Proposition 3. *Suppose that there exists a compact set $\bar{Z}$ of $R^n \times R$ such that: (i) S is nonempty and is contained in $\bar{Z}$; (ii) all conditions in Proposition 2 are satisfied on the set $\bar{Z}$; (iii) w.p.1 for N large enough the set $\bar{S}_N$ is nonempty and $\bar{S}_N \subset \bar{Z}$; (iv) If $z_N = (x_N, u_N)$ be feasible to (8.13) and $z_N \to z$ w.p.1, then z is a feasible point of (8.11); (v) For some point $z \in S$, there exists a sequence of $\{z_N\}$ being feasible to (8.13) such $z_N \to z$ w.p.1. Then $\bar{v}_N \to v^*$ w.p.1 as $N \to \infty$.*

Proof. For simplicity, let g and Z denote the objective function and the feasible sets of the original problem (8.11), g_N and Z_N denote the objective and the feasible set of the SAA program (8.13). By virtue of the indicator function [17], $\delta(\cdot|\theta)$, where θ denotes a set of any finite dimensional vector space, then the original problem (8.11) and its SAA problem (8.13) can be rewritten as:

$$\min\{g(z) + \delta(z|Z)|z \in R^n \times R\}$$

and

$$\min\{g_N(z) + \delta(z|Z_N)|z \in R^n \times R\},$$

respectively. Thus, we only need to investigate the convergence of the above two unconstrained minimization problems. By Proposition 2, it is easy to see that $g_N(z)$ converges to $g(z)$ w.p.1 uniformly on $\bar{Z}$. Then, the result follows directly with help of ([22], Proposition 6 and Remark 8).

Note that the conditions in (8.3) can be satisfied by imposing some appropriate properties on the underlying functions such as convexity or continuity, together with some constraint qualifications such as the Slater condition. But, this is beyond the scope of this chapter. The reader is referred to [22] for a detailed discussion. In many cases, the underlying functions f, h_l, $l = 1, \ldots, k$, are all smooth. However, the term $[f(x,y) - u]^+$ is nonsmooth, which leads to the nonsmoothness of problem (8.11) and its SAA counterpart (8.13) generally. To overcome the nonsmoothness, we use some well-developed smoothing techniques. First, note that

$$[f(x,y) - u]^+ = -[-(f(x,y) - u)]^- = -\min\{0, u - f(x,y)\}.$$

So, we may replace the term $[f(x,y) - u]^+$ in (8.11) by using the min function $\min\{a,b\}$. However, this function is nonsmooth as well, so we consider the following smoothed counterpart of $\min\{a,b\}$:

$$\psi(a,b,\mu) = -\frac{1}{2}\left(\sqrt{(a-b)^2 + \mu^2} - (a+b)\right),$$

which is the Chen–Harker–Kanzow–Smale (CHKS) smoothing function [12, 16, 29] with $\mu \neq 0$ as the *smoothing parameter*, which is driven to zero in computation. Clearly, ψ is continuously differentiable everywhere except at $\mu = 0$ and $\psi(a,b,0) = \min\{a,b\}$. The smoothing function for $\min\{0, u - f(x,y)\}$ is then as follows:

$$\psi(0, u - f(x,y), \mu) = -\frac{1}{2}\left(\sqrt{(u - f(x,y))^2 + \mu^2} - u + f(x,y)\right).$$

In practice, we choose a sequence $\{\mu^l\}$ with positive small values of μ^l and letting $\mu^l \to 0$ as $l \to \infty$. For any given $\mu \neq 0$, we derive the smoothing version of (8.11) as follows.

$$
\begin{aligned}
&\min(1-\lambda)\mathrm{E}\left[f(x,y)\right]+\lambda\left(u+\frac{1}{2(1-\alpha)}\mathrm{E}\left[\begin{array}{l}\sqrt{(u-f(x,y))^2+\mu^2}-u+\\ f(x,y)\end{array}\right]\right)\\
&s.t.\quad \mathrm{E}\left[h_l(x,y)\right]=0\,,\quad l=1,2,\ldots,j\,, \qquad (8.14)\\
&\mathrm{E}\left[h_l(x,y)\right]\le 0\,,\quad l=j+1,\ldots,k\,,\\
&u+\frac{1}{2(1-\alpha)}\mathrm{E}\left[\left(\sqrt{(u-f(x,y))^2+u^2}-u+f(x,y)\right)\right]\le\beta\,,\\
&(x,u)\in X\times R\,.
\end{aligned}
$$

Similarly, the corresponding smoothing SAA program is as follows.

$$
\begin{aligned}
&\min\frac{1-\lambda}{N}\sum_{i=1}^{N}f(x,y^i)\\
&\qquad+\lambda\left(u+\frac{1}{2N(1-\alpha)}\sum_{i=1}^{N}\left(\sqrt{(u-f(x,y^i))^2+\mu^2}-u+f(x,y^i)\right)\right)\\
&s.t.\quad \frac{1}{N}\sum_{i=1}^{N}h_l(x,y^i)=0\,,\quad l=1,2,\ldots,j\,, \qquad (8.15)\\
&\frac{1}{N}\sum_{i=1}^{N}h_l(x,y^i)\le 0\,,\quad l=j+1,\ldots,k\,,\\
&u+\frac{1}{2N(1-\alpha)}\sum_{i=1}^{N}\left(\sqrt{(u-f(x,y^i))^2+\mu^2}-u+f(x,y^i)\right)\le\beta\,,\\
&(x,u)\in X\times R\,.
\end{aligned}
$$

Thus, to derive an approximate solution of (8.10), we can solve the SAA smoothing problem (8.15) using some well developed optimization approaches.

8.4 An Illustration of the Stochastic Model

We now illustrate the stochastic model and the SAA method discussed above by considering a supply chain consisting of two suppliers, one contract manufacturer, and two distributors. The manufacturer needs to procure raw materials from the suppliers and sell the completed products to the distributors. Denote by p_1, p_2 the prices of raw materials sold to the manufacturer and by q_1 and q_2 the prices of the product sold to the distributors. Denote by c the unit shipping cost of raw materials paid by the manufacturer and ρ the raw material loss rate. Let α denote the confidence level and β denote the loss tolerance.

Let $x_1$and x_2 denote the quantities of raw materials purchased by the manufacturer, x_3 and $x_4$the quantities of products sold to the distributors. Clearly, the total quantity, $x_3 + x_4$, of completed products sold to the distributors cannot exceed the difference of the total quantity of raw materials and the total loss of raw materials, $(1-\rho)(x_1+x_2)$. So, we have the following constraint:

$$(1-\rho)(x_1+x_2) - x_3 - x_4 \geq 0\,.$$

In addition, the capacities for these two suppliers are set as follows:

$$0 \leq x_1 \leq 40\,, \quad 0 \leq x_2 \leq 40\,, \text{ and } x_1 + x_2 \leq 60\,.$$

For simplicity, we consider two types of external risk involved in the chain. Specifically, let y_1 denote the random variable concerning the risk (loss) of a supplier defaulting and y_2 the random variable concerning the risk (loss) of no demand from the distributors. The loss function is given by

$$f(x,y) := p_1x_1 + p_2x_2 + c(x_1+x_2) - q_1x_3 - q_2x_4 + w_1y_1 + w_2y_2\,,$$

where w_i represents the risk severity factor with $w_i \geq 0$ and $w_1 + w_2 = 1$. In this example, the underlying loss is irrespective of the internal factors related to the quantities of raw materials or/and the completed products. In fact, the loss function can be written separately as $f(x,y) = f_1(x) + f_2(y)$ where $f_1(x)$ represents the normal costs of the manufacturer while $f_2(y) = w^{\mathrm{T}}y$ denotes the loss due to the risks only. Note that in general the loss function might not be separable in the decision variable x and the random variable y. Note also that the loss function f can be nonlinear and even nonsmooth in some situations [11, 13].

Based on the previous arguments, the corresponding SAA program and the smoothing SAA program can then be written as follows:

$$\begin{aligned}
&\min \frac{1-\lambda}{N}\sum_{i=1}^{N} f(x,y^i) + \lambda\left(u + \frac{1}{N(1-\alpha)}\sum_{i=1}^{N}[f(x,y^i)-u]^+\right)\\
&s.t.\ (1-\rho)(x_1+x_2) - x_3 - x_4 \geq 0\,,\\
&x_1 + x_2 \leq 60\,, \quad 0 \leq x_i \leq 40\,, \quad i = 1,2\,, \quad x_3, x_4 \geq 0\,,\\
&u + \frac{1}{N(1-\alpha)}\sum_{i=1}^{N}[f(x,y^i)-u]^+ \leq \beta\,,\\
&x \in R^4\,, \quad u \in R\,,
\end{aligned}$$

and

$$
\begin{aligned}
&\min \tfrac{1-\lambda}{N} \sum_{i=1}^{N} f(x,y^i)+\\
&\lambda\left(u+\tfrac{1}{2N(1-\alpha)}\sum_{i=1}^{N}\left(\sqrt{(u-f(x,y^i))^2+\mu^2}-u+f(x,y^i)\right)\right)\\
&s.t.\ (1-\rho)(x_1+x_2)-x_3-x_4 \geq 0\,,\\
&x_1+x_2 \leq 60\,,\quad 0 \leq x_i \leq 40\,,\quad i=1,2\,,\quad x_3,x_4 \geq 0\,,\\
&u+\frac{1}{2N(1-\alpha)}\sum_{i=1}^{N}\left(\sqrt{(u-f(x,y^i))^2+\mu^2}-u+f(x,y^i)\right) \leq \beta\,,\\
&x \in R^4, u \in R\,.
\end{aligned}
$$

In the numerical test, we set $\lambda = 0.2$, the values for p_1, p_2 and q_1, q_2, c are set to be 2.1, 2.3, 5.5, 6.0, 0.75 per unit of raw materials or completed products, respectively. Set $\beta = 0.001$, $\rho = 0.05$, $w_1 = 0.7$, $w_2 = 0.3$. For simplicity, the random vector $y = (y_1, y_2)^{\mathrm{T}}$ is assumed to follow the uniform distribution on $[10, 40]$.

We conduct the test using the Matlab built-in solver *fmincon* with different values of the smoothing parameter μ and sample sizes N, M and N'. Let $(\bar{x}_N, \bar{u}_N)$ denote the solution of the corresponding SAA program. Note that the numbers of samples for deriving the lower and upper bounds are $N \times M$ and N', respectively. The results are displayed in Table 8.1.

Table 8.1 Numerical results for a supply chain risk management problem

α	μ	N	M	N'	$\bar{x}_N$	$\bar{u}_N$
0.95	10^{-4}	1500	100	4000	(40, 20, 0, 57)	−128.6837
0.95	10^{-4}	2000	100	5000	(40, 20, 0, 57)	−128.5688
0.95	10^{-4}	2500	100	6000	(40, 20, 0, 57)	−128.6501
0.98	10^{-6}	1000	100	4000	(40, 20, 0, 57)	−128.1244
0.98	10^{-6}	2000	100	5000	(40, 20, 0, 57)	−127.9133
0.98	10^{-6}	2500	100	6000	(40, 20, 0, 57)	−128.0078

8.5 A Wine Supply Chain Problem

In this section, we consider a wine industry logistics problem, which is modified from Yu and Li [26]. A wine company is formulating its production, inventory, and transportation plan. This company owns three bottling plants located in E, F, and G, and three distribution warehouses located in three different countries (or cities)

L, M, and N, respectively. Uniform-quality wine in bulk (raw material) is supplied from four wineries located in A, B, C, and D.

For simplicity, without considering other market behaviors (e. g. novel promotion, marketing strategies of competitors, and market-share effect in different markets), each market demand merely depends on the local economic conditions. Assume that the future economy is either boom, good, fair, or poor, i. e., four situations with associated probabilities of 0.45, 0.25, 0.17, or 0.13, respectively. The unit production costs and market demands under each situation are listed in Table 8.2.

Table 8.2 Characteristics of the problem

Future	Demands			Unit product cost			Likelihood of each
Economic situation	L	M	N	E	F	G	economic situation
Boom	400	188	200	755	650	700	0.45
Good	350	161	185	700	600	650	0.25
Fair	280	150	160	675	580	620	0.17
Poor	240	143	130	650	570	600	0.13

Let i denote the index of four wineries, j the index of the three distribution centers, and k the index of three bottling plants. We call each possible economic situation a "scenario", denoted by s. For $i = 1, 2, 3, 4$, $j = 1, 2, 3$, $k = 1, 2, 3$, and each scenario s, we define the following notation:

- x_{ik} – the amount of bulk wine to be shipped from winery i to bottling plant k,
- y_k – the amount of bulk wine to be bottled in bottling plant k,
- z_{kj} – the amount of bottled wine shipped from bottling plant k to distribution center j,
- u_k – the stock of bulk wine in bottling plant k carried from previous weeks,
- v_k – the stock of bottled wine in bottling plant k carried from previous weeks,
- c_s^j – the uncertain unit cost, depending on each scenario s, for bottling wine in bottling plant j,
- D_s^j – the uncertain market demand, depending on each scenario s, from the distribution center j.

In the problem, the random factors involved are uncertain unit costs/prices from bottling plants and the random demands from distributors. Let

$$\xi := (D^1, D^2, D^3, c^1, c^2, c^3)$$

denote the underlying random vector. For convenience, we use the index set {1, 2, 3, 4} to denote the four observations/scenarios of the future economy situation {Boom, Good, Fair, Poor}, respectively. For example, the index corresponds to "Boom". Then, $\xi(s) := (D_s^1, D_s^2, D_s^3, c^1, c^2, c^3)$, $s = 1, 2, 3, 4$, are all possible values of ξ, as per Table 8.2. We denote the decision variables by w, where $w := (x, y, z, u, v)$, $x := (x_{11}, \ldots, x_{43})$, $y := (y_1, y_2, y_3)$, $z := (z_{11}, \ldots, z_{33})$, $u := (u_1, u_2, u_3)$, and $v := (v_1, v_2, v_3)$.

We assume that (65.6, 155.5, 64.7, 62, 150.5, 60.1, 84, 174.5, 88.4, 110.5, 100.5, 109.3) are the unit costs of transporting bulk wine from each winery A, B, C, and D to each bottling plant E, F, and G, respectively. Also, we assume that (200.5, 300.5, 699.5, 693, 533, 362, 164.9, 306.4, 598.2) are the unit costs of transporting bottled wine from each bottling plant E, F, and G to each distribution center L, M, and N, respectively. We denote by (75, 60, 69) the unit inventory costs for bulk wine in bottling plant E, F, and G, respectively, and by (125, 100, 108) the unit inventory costs for bottled wine in bottling plant E, F, and G, respectively.

In addition, we denote by ρ_i, $i = 1, 2, 3, 4$, the maximum amount of bulk wine that can be shipped from winery i to the bottling plants; denote by υ_k, $k = 1, 2, 3$, the maximum production capacity of bottled wine for each bottling plant k; denote by π_k, $k = 1, 2, 3$, the maximum storage space for the bulk wine in plant k; denote by θ_k, $k = 1, 2, 3$, the maximum storage spaces for bulk wine in plant k. Here, we set $\rho = (300, 150, 200, 150)$, $y = (340, 260, 300)$, $\pi = (150, 100, 120)$, $\theta = (50, 50, 50)$.

The total cost f consists of the transportation cost f_T, production cost f_p, and inventory cost f_I. For each realization $\xi(s)$, the above three cost functions are given as follows:

$$\begin{aligned} f_T(w, \xi(s)) = {} & 65.6x_{11} + 155.5x_{12} + 64.7x_{13} + 62x_{21} + 150.5x_{22} + 60.1x_{23} \\ & + 84x_{31} + 174.5x_{32} + 88.4x_{33} + 110.5x_{41} + 100.5x_{42} + 109.3x_{43} \\ & + 200.5z_{11} + 300.5z_{12} + 699.5z_{13} + 693z_{21} + 533z_{22} \\ & + 362z_{23} + 164.9z_{31} + 306.4z_{32} + 598.2z_{33}, \end{aligned}$$

$$f_P(w, \xi(s)) = c_s^1 y_1 + c_s^2 y_2 + c_s^3 y_3,$$

and

$$\begin{aligned} f_I(w, \xi(s)) = {} & 75(x11 + x21 + x31 + x41 - y1 + u1) + 60(x12 + x22 \\ & + x32 + x42 - y2 + u2) + 69(x13 + x23 + x33 + x43 \\ & - y3 + u3) + 125(y1 + v1 - z11 - z12 - z13) + 100(y2 + v2 \\ & - z21 - z22 - z23) + 108(y3 + v3 - z31 - z32 - z33). \end{aligned}$$

Consequently, for each scenario s, the total cost function is as follows.

$$f(w, \xi(s)) = f_T(w, \xi(s)) + f_P(w, \xi(s)) + f_I(w, \xi(s)).$$

Note that f is actually a function involved with a random factor ξ. The stochastic model under consideration is:

$$\begin{aligned} & \min \ \mathrm{CVaR}_\alpha(w) \\ & s.t. \quad x_{i1} + x_{i2} + x_{i3} \le \rho_i, \quad i = 1, 2, 3, 4, \\ & \qquad\ y_k \le \upsilon_k, \quad k = 1, 2, 3, \end{aligned} \tag{8.16}$$

$$
\begin{aligned}
&0 \le y_k + v_k - z_{k1} - z_{k2} - z_{k3} \le \theta_k\,, \quad k = 1, 2, 3\,,\\
&0 \le z_{1j} + z_{2j} + z_{3j} - D_s^j\,, \text{ for each } s\,,\, j = 1, 2, 3\,,\\
&x_{ik}, y_k, z_{kj}, u_k, v_k \ge 0\,, \quad i = 1, 2, 3, 4\,, \quad k = 1, 2, 3\,, \quad j = 1, 2, 3\,.
\end{aligned}
$$

Based on the previous discussions, the above problem can be formulated equivalently as:

$$
\begin{aligned}
&\min \gamma + \frac{1}{1-\alpha} \mathrm{E}\,[f(w,\xi) - \gamma]^+\\
&\quad s.t. \quad x_{i1} + x_{i2} + x_{i3} \le \rho_i\,, \quad i = 1, 2, 3, 4\,,\\
&\qquad\quad y_k \le v_k\,, \quad k = 1, 2, 3\,,\\
&0 \le x_{1k} + x_{2k} + x_{3k} + x_{4k} + u_k - y_k \le \pi_k\,, \quad k = 1, 2, 3\,,\\
&0 \le y_k + v_k - z_{k1} - z_{k2} - z_{k3} \le \theta_k\,, \quad k = 1, 2, 3\,,\\
&0 \le z_{1j} + z_{2j} + z_{3j} - D_s^j\,, \text{ for each } s,\, j = 1, 2, 3\,,\\
&\gamma \in R\,, \quad x_{ik}, y_k, z_{kj}, u_k, v_k \ge 0\,, \quad i = 1, 2, 3, 4\,, \quad k = 1, 2, 3\,, \quad j = 1, 2, 3\,.
\end{aligned}
\tag{8.17}
$$

Using the CHKS smoothing function as discussed in Sect. 8.3, we solve the following smoothed discretized problem of (8.16):

$$
\begin{aligned}
&\min \gamma + \frac{1}{2(1-\alpha)} \sum_{s=1}^{4} p_s \left[\sqrt{4\varepsilon^2 + (f(w,\xi(s)) - \gamma)^2} + f(w,\xi(s)) - \gamma \right]\\
&s.t. \quad x_{i1} + x_{i2} + x_{i3} \le \rho_i\,, \quad i = 1, 2, 3, 4\,,\\
&\qquad y_k \le v_k\,, \quad k = 1, 2, 3\,,\\
&\qquad 0 \le x_{1k} + x_{2k} + x_{3k} + x_{4k} + u_k - y_k \le \pi_k\,, \quad k = 1, 2, 3\,,\\
&\qquad 0 \le y_k + v_k - z_{k1} - z_{k2} - z_{k3} \le \theta_k\,, \quad k = 1, 2, 3\,,\\
&\qquad 0 \le z_{1j} + z_{2j} + z_{3j} - D_s^j\,, \quad \text{for each } s, j = 1, 2, 3\,,\\
&\qquad \gamma \in R\,,\ x_{ik}, y_k, z_{kj}, u_k, v_k \ge 0\,,\ i = 1, 2, 3, 4\,,\ k = 1, 2, 3\,,\ j = 1, 2, 3,
\end{aligned}
\tag{8.18}
$$

where $\varepsilon > 0$.

We have carried out numerical tests on the problem and report some numerical results. Again, the tests are carried out by implementing mathematical programming codes in MATLAB 6.5 installed in a PC with the Windows XP operating system. We used the MATLAB built-in solver *fmincon* for solving the associated problems.

We report the amounts of bulk wine to be shipped and bottled x, y; the amount of bottled wine shipped z; and the corresponding total cost, the transportation cost, the production cost, the inventory holding cost, the minimum conditional value at risk CVaR_α, the value at risk of the total cost VaR_α.

Table 8.3 Numerical results of the smoothing method for wine supply chain problem

$\varepsilon = 10^{-4}, \alpha = 0.95$	Result
x	(0.0000.0.0000.25.3640.29.9288.32.5259.25.3476, 29.9288. 32.5259. 25.3476. 29.9288. 32.5259. 25.3476)
y	(119.7384.117.6691.123.3638)
z	(136.5054.64.0625.68.1631.135.9777.63.5347, 67.6353. 137.4613. 65.0183. 69.1189)
Total cost	603 490
Transportation cost	360 510
Production cost	239 410
Inventory holding cost	3 570
CVaRα	1
VaRα	0.5000

In Table 8.3, we set the smoothing parameter $\varepsilon = 10^{-4}$ and the confidence level $\alpha = 0.95$. In the numerical test, we have also chosen different values for the smoothing parameter and the confidence level, such as $\varepsilon = 10^{-5}$, 10^{-6}, 10^{-7}, and $\alpha = 0.9$, 0.96. We also notice that, for the problem under consideration, the numerical results stated in Table 8.3 do not change greatly with small changes in the values of the confidence level and the smoothing parameter.

In this example, we have attempted to highlight that risk can and must be measured in every supply chain. In the case of the wine supply chain, we show that VaR does not cover risks of interest (CVaR $\neq$ VaR). Further, having a numerical value attached to either VaR or CVaR is helpful to mangers and decision makers on providing a simple but comprehensive statement on the exposure to the supply chain risk on an aggregate level. This can then help mangers to further decide on what actions are needed to mitigate such risks, for instance, changing the choice of α to get a lower quantitation risk score, which could then translate to keeping more wine or increasing higher inventory and transportation costs.

8.6 Conclusion

In this chapter, we have established a stochastic model for supply chain risk management using a measure of risk, the conditional value at risk. For the case where the random variable follows a continuous distribution, we introduced a sample average approximation approach in combination with some smoothing techniques to solve the underlying stochastic problem. This makes each iterative subproblem a smooth and convex optimization problem. We illustrate the established model by considering a small supply chain management example. Finally, a wine supply chain under uncertain demand with a discrete distribution is discussed.

This research, as applied to the supply chain area, is timely and useful given the recent spate of fuel cost disruptions to the supply chain and the current economic crisis, which ultimately affects production and incurs a need for better management control of the overall supply chain of supplies, manufacturers, and distributors. Our proposed technique is especially in dealing with problems associated with uncertainties in distributions, where the traditional objective of profit maximization is still desired while constraining the risk of a false production or untrue demand.

Acknowledgements We would like to thank the referees for their constructive suggestions, which helped to improve the presentation of this chapter.

References

1. F. ANDERSSON, H. MAUSSER, D. ROSEN, AND S. URYASEV. *Credit risk optimization with conditional Value-at-Risk criterion*, Mathematical Programming, 89(2001), pp. 273–291
2. P. ARTZNER, F. DELBAEN, J.M. EBER, AND D. HEATH, *Coherent measures of risk*, Mathematical Finance, 9(1999), pp. 203–228
3. F. BASTIN, C. CIRILLO, AND P.L. TOINT, *Convergence theory for nonconvex stochastic programming with an application to mixed logit*, Mathematical Programming, 108(2006), pp. 207–234
4. E. BOGENTOFT, H.E. ROMEIJN, AND S. URYASEV, *Asset/liability management for pension funds using CVaR constraints*, The Journal of Risk Finance, 3(2001), pp. 57–71
5. M. GOH, J. LIM, AND F. MENG, *A stochastic model for risk management in global supply chain networks*, European Journal of Operational Research, 182 (2007) 164–173
6. G. GÜRKAN, A.Y. ÖZGE, AND S.M. ROBINSON, *Sample-path solution of stochastic variational inequalities*, Mathematical Programming, 84(1999), pp. 313–333
7. W. HURLIMANN, *Conditional Value-at-Risk bounds for compound poisson risks and a normal approximation*, Journal of Applied Mathematics, 3(2003), pp. 141–153
8. P. JORION, *Value At Risk: the New Benchmark for Controlling Market Risk*, McGraw-Hill, New York, 1997; McGraw-Hill International Edition, 2001
9. P. KROKHMAL, J. PALMQUIST, AND S. URYASEV, *Portfolio optimization with conditional Value-At-Risk objective and constraints*, The Journal of Risk, 4(2002), pp. 43–68
10. N. LARSEN, H. MAUSSER, AND S. URYASEV, *Algorithms for optimization of value at risk*, in P. Pardalos and V.K. Tsitsiringos, editors, Financial Engineering, E-Commerce and Supply Chain. Kluwer Academic Publishers, Norwell, 2002, pp. 129–157
11. F. MENG, J. SUN, AND M. GOH, *A smoothing sample average approximation method for CVaR optimization*, working paper, The Logistics Institute – Asia Pacific, School of Business, National University of Singapore, May 2008
12. F. MENG AND H. XU, *A regularized sample average approximation method for stochastic mathematical programs with nonsmooth equality constraints*, SIAM Journal on Optimization, 17(2006), pp. 891–919
13. J.-S. PANG AND B.F. HOBBS, *Nash-Cournot equilibria in electric power markets with piecewise linear demand functions and joint constraints*, Operations Research, 55 (2007), pp. 113–127
14. J.-S. PANG AND S. LEYFFER, *On the global minimization of the Value-at-Risk*, Optimization Methods and Software, 19 (2004), pp. 611–631
15. G. PFLUG, *Some remarks on the Value-at-Risk and the conditional Value-at-Risk*, in Probabilistic Constrained Optimization: Methodology and Applications, S. Uryasev (ed.) Kluwer Academic Publishers, 2000, pp. 1–10
16. L. QI, D. SUN, AND G. ZHOU, *A new look at smoothing Newton methods for nonlinear complementarity problems and box constrained variational inequalities*, Mathematical Programming, 87(2000), pp. 1–35

17. R.T. ROCKAFELLAR, *Convex Analysis*, Princeton University, 1970
18. R.T. ROCKAFELLAR AND S. URYASEV, *Conditional Value-at-Risk for general loss distributions*, Journal of Banking & Finance, 26(2002), pp. 1443–1471
19. R.T. ROCKAFELLAR AND S. URYASEV, *Optimization of conditional Value-at-Risk*, Journal of Risk, 2(2000), pp. 21–41
20. A. RUSZCZYNSKI AND A. SHAPIRO, *Stochastic Programming*, Handbooks in Operations Research and Management Science, Volume 10, North-Holland, Amsterdam, 2003
21. T. SANTOSO, S. AHMED, M. GOETSCHALCKX, AND A. SHAPIRO, *A stochastic programming approach for supply chain network design under uncertainty*, European Journal of Operational Research, 67(2005), pp. 96–115
22. A. SHAPIRO, *Monte Carlo sampling methods.* In A. Ruszczynski, A. Shapiro (eds.), Stochastic Programming, Handbooks in OR & MS, Vol. 10, North-Holland Publishing Company, Amsterdam, 2003
23. A. SHAPIRO, *Stochastic mathematical programs with equilibrium constraints*, Journal of Optimization Theory and Application, 128 (2006), pp. 223–243
24. D. TASCHE, *Expected shortfall and beyond*, Journal of Banking and Finance, 6(2002), pp. 1519–1533
25. C.S. TANG, *Perspectives in Supply Chain Risk Management*, International Journal of Production Economics, 103 (2006), pp. 451–488
26. C.-S. YU AND H. LI, *A robust optimization model for stochastic logistic problems*, International Journal of Production Economics, 64(2000), pp. 385–397
27. S. URYASEV, *Introduction to the theory of probabilistic functions and percentiles (Value-at-Risk)*, Probabilistic Constrained Optimization: Methodology and Applications, S.P. Uryasev (ed.), Kluwer Academic Publishers, 2000, pp. 1–25
28. S. URYASEV, *Conditional Value-at-Risk: optimization algorithms and applications*, Financial Engineering News, 14, 2000
29. H. XU AND F. MENG, *Convergence analysis of sample average approximation methods for a class of stochastic mathematical programs with equality constraints*, Mathematics of Operations Research, 32(2007), pp. 648–668

Chapter 9
Risk Intermediation in Supply Chains

Ying-Ju Chen and Sridhar Seshadri

Abstract In some supply chains, retailers are relatively small and averse to taking risk. In such a situation, traditional methods of contracting, that typically assume risk neutrality on retailers, might not suffice to maximize the seller's/distributor's expected profit. We present tools for analyzing and solving such a problem from the viewpoint of a (risk-neutral) seller/distributor. We present two types of models that can be used to create contracts, one set in a discrete setting and the other in a continuous setting. In both settings, individual retailers are characterized with different degrees of risk aversion.

We first explain in the discrete setting how the varying degrees of risk aversion present hurdle for the design of a uniform contract for all retailers. We then show how to mitigate the problem using ideas from the theory of mechanism design. We offer a simple solution to the contract design problem and show how it can be easily implemented. We next show that in the continuous setting in which the distribution of retailers is continuous, which could be viewed as a limiting case of the discrete setting, the contract design problem actually simplifies. In this continuous setting, we show that it becomes relatively easy to design contracts and establish their optimality from the seller's/distributor's viewpoint. We conclude the chapter with a summary of problems that are still open in this area.

9.1 Introduction

We consider a single period model in which multiple risk-averse retailers purchase a single product from a common vendor. We assume that the retailers operate in identical and independent markets. The retailers face uncertain customer demand and accordingly make their purchase order quantity decisions in order to maximize their expected utility. The vendor offers the same supply contract to each retailer. The terms of the contract offered to the retailers are similar to the ones found in the classical newsvendor problem. Under this contract, each retailer purchases a certain

T. Wu, J. Blackhurst (eds.), *Managing Supply Chain Risk and Vulnerability*

quantity at a regular purchase price. If the realized demand is greater than the quantity ordered then the retailer has the option to purchase the units that are short at an emergency purchase price which is higher than the regular price. If the demand is less than the order quantity then the retailer has the option to return the leftover inventory at a salvage price that is lower than the regular price. (This contract is referred to in what follows as the original newsvendor contract (ONC).) The retailers are price takers and sell the product at the same fixed price. The problem of deciding upon the quantity to order from the vendor is similar to the classic newsvendor problem, except that due to risk aversion, each retailer maximizes his expected utility rather than his expected profit. The problem faced by a newsvendor is known as the "risk-averse newsvendor problem".

This problem has been well studied in the literature. In particular is well known that the risk-averse retailer's order quantity (i. e., the one that maximizes his expected utility) will be smaller than the order quantity that maximizes his expected profit (see Horowitz, 1970; Baron, 1973; Eeckhoudt et al., 1995). Obviously, the reduction in the order quantity of the retailer leads to a lower expected profit (for the retailer) compared to the expected profit obtained under the profit maximizing order quantity. Eeckhoudt et al. give examples in which risk averse retailers will order nothing due to high demand uncertainty. Therefore, risk aversion of the retailers has been portrayed in the literature as leading to the loss of efficiency in supply chains. (We use the term "efficiency" to refer to the combined expected profit of the seller and the retailer. In general, this term refers to the total expected profit of all participants in a supply chain.)

In this chapter we show not only that this loss of efficiency can be eliminated through risk reducing pricing contracts but also that any risk neutral intermediary will find it beneficial to offer such risk reducing contracts to the retailers. In our model, the intermediary is referred to as the distributor[1] and purchases the goods as per the terms of the ONC from the vendor. In turn the distributor offers the goods to the retailers on contract terms that are less risky from the retailers' viewpoint, namely: We propose that, as opposed to the ONC, under the risk reducing contracts offered by the distributor to the retailers, the emergency purchase and the salvage prices should be set equal to the regular purchase price, and that in addition a fixed payment should be made by the distributor to the retailer. Therefore, a retailer's payoff consists of a fixed component (independent of the demand) and a variable component that increases linearly with the realized demand. Consequently, as the retailer's payoff depends only upon the demand, the retailer is indifferent to the order quantity decision and is content to relegate the responsibility of determining an order quantity to the distributor. The distributor makes the order quantity decision fully aware that he has to satisfy all the demand faced by the retailer. Thus the distributor bears the cost if necessary of buying the product at the emergency purchase cost and

[1] The distributor can be an independent firm, the vendor, or one of the retailers. For the sake of clarity we will refer to the intermediary as the Distributor, and the risk averse players facing uncertain demand as the Retailers. The analysis, though, is valid for any two levels in a vertical marketing channel, where the lower level facing uncertain demand is risk averse and the upper level is risk neutral (or less risk averse).

also if necessary of disposing any unsold product at the salvage price. We prove that by performing this function of "(demand) risk intermediation" the distributor raises the retailers' order quantities such that the maximum efficiency is obtained. The key theorem in this chapter is to establish that the contracts offered to the retailer not only maximize the efficiency in the supply chain but are also optimal from the distributor's viewpoint.

Contracts similar to the ones proposed in this chapter are being adopted within the context of vendor managed inventory (VMI) programs. In many VMI programs the vendors make the inventory decisions on behalf of the retailers and also bear the risks and costs associated with these decisions (Andel, 1996). In addition to the contracts found in VMI programs, we have observed several supply contracts, for example in the publishing, cosmetics, computers, apparel and grocery industries, that transfer the demand risk from the buyer to the vendor. In the publishing industry (Carvajal, 1998) the retail outlets return their unsold magazines to the distributor at their purchase price and get additional shipments if they run out (such contracts were first introduced in the depression era). The terms of this contract are instrumental in persuading small retail merchants, who are averse to risk, to stock sufficient quantity of a wide variety of magazines.

Moses and Seshadri (2000) describe the incentives used by the manufacturers in the cosmetics industry to increase the stocking level at department stores. The incentives in the cosmetic industry comprise of liberal return policies and the sharing of inventory holding costs between the producer and the department stores. In the personal computer industry (Kirkpatrick, 1997), which is an industry plagued by steep price depreciation, the standard practice of vendors is to offer price guarantees to their VAR's (value added resellers) and hence the vendors absorb the risk of price erosion of their products during the period that they are held in the retailers' inventory. Similarly apparel retailers, who as an industry are facing increasing markdowns, are pressurizing their vendors to offer margin guarantees (Bird and Bounds, 1997), i. e., the vendors are expected to absorb the markdown risk faced by the retailers in case the goods have to be disposed at "sale" prices. The grocery retailers' response to increasing inventory risk due to the proliferation of SKU's (Lucas, 1996) is to charge a fixed slotting fees (similar to the fixed fee proposed by us), ranging from $ 5000 to $ 20,000 per year per SKU (stock keeping unit), from the manufacturers/distributors irrespective of the sale volume of the items. The slotting fee is simply a rent charged for use of the shelf space. Therefore, we see a trend in the industry where simple price discounting contracts that were previously offered by the vendors (to induce the retailers to purchase a larger quantity) are being substituted by relatively sophisticated contracts that are designed to transfer the demand risk from the retailers to their vendors. The economic justification for such contracts is not complete unless as done in this chapter, the impact of risk reduction for risk-averse retailers is carefully accounted for, and the optimal form of incentives (i. e., contracts) is established taking into account the nature of the risk and the retailer's aversion to risk.

Our work shows how supply contracts can play a crucial role in the reduction of financial risk resulting from demand uncertainty. We model the role of incentives (embodied in pricing contracts) in supply chains where uncertainty leads to inefficient decision making. Instead of entirely focusing on the risk-averse retailers' order quantity decisions (as done in the entire literature dealing with the risk averse newsvendor problem) the focus is on *optimal* mechanisms that can be used to influence the decisions of the retailers. In our model, we assume that the retailers operate in identical and independent markets and the problem setting for each retailer is the single period newsvendor problem. This problem of designing pricing contracts under demand uncertainty and risk aversion is complicated because of the following reasons:

(1) Retailers might differ with regard to their aversion to risk. Therefore, different retailers may derive different expected utility from the same contract.
(2) The distributor does not know how risk averse any particular retailer is and only knows the *distribution* of risk aversion among the retailers.
(3) The laws against price discrimination prohibit the distributor from offering different contracts selectively to different retailers, i. e., any contract that the distributor decides to offer to a specific retailer, has to be offered to all the retailers.

We begin our analysis by showing that in general it is not in the distributor's best interest to offer the *same* contract to all retailers. In other words, the distributor does not in general maximize either his expected profit or the efficiency, by simply performing the functions of buying and reselling the product to all retailers on the *same* terms. A *menu* of contracts is necessary to maximize the expected profit of the distributor, and in fact the fundamental contribution of this chapter is to exactly characterize the *menu* that the distributor should offer in order to maximize his expected profit. This menu rather interestingly maximizes the efficiency as well. Every contract in the menu has two parameters, a fixed payment that the distributor makes to the retailer, and a unit price that the distributor charges the retailer for every unit sold by the retailer. (The choice of this form of each contract in the menu is neither random nor accidental. We prove that these contract terms create stochastic payoffs for the risk-averse retailer that dominates the payoffs under the terms of the ONC in a strong sense.)

We emphasize that the menu does not depend on the distributor's knowledge of the degree of risk aversion of each and every retailer, but only upon the knowledge of the distribution of risk aversion over the ensemble of retailers. The menu of contracts is either continuously or discretely parameterized by the fixed side payment and the selling price depending upon whether the distribution of risk aversion is a discrete one or a continuous one. Each retailer selects a contract from the menu, choosing the one that provides him the highest expected utility. The menu of contracts derived by us has the following special properties:

- In order to maximize his expected profit, the distributor should offer a menu of contracts so tailored that every (risk averse) retailer selects a unique contract from the menu. This fact is not readily discernible because the distributor could choose to create a menu such that some retailer finds nothing attractive in the menu.

- Each retailer chooses a contract from the menu and by doing so obtains as much or higher expected utility (EU), as compared to the EU from the original newsvendor contract (ONC). Therefore, all retailers are as well off or better off with the entry of the distributor in the supply chain.
- The contracts in the menu are independent of the order quantity. The retailer is willing to relegate the ordering decision to the distributor. Furthermore, the order quantity stipulated by the distributor is the expected value (EV) maximizing solution to the newsvendor problem. Therefore, the mutually beneficial (beneficial to retailers and the distributor) menu of contracts is also instrumental in maximizing efficiency.
- The menu of contracts is such that retailers who are less risk averse, prefer the contracts with higher expected profit and utility (and consequently bear higher risk).
- The menu always contains a risk free (fixed payment) contract. This contract is selected by a subset of retailers, namely those that are the most risk averse. Under fairly general conditions, the subset of retailers who select this risk free contract is independent of the product characteristics, and dependent only upon the retailers' attitude towards risk. This shows that the fraction of retailer population that will face no demand risk (and in effect are owned by the distributor) does not depend on the product but on the distribution of risk aversion.

It is common knowledge how in the last two decades the concept of risk intermediation has been used to create not only novel investment and insurance products but also a global marketplace for such products and services. A large number of firms now offer a menu of products with different risk-return choice to customers worldwide. Viewed in this light, the existence of a similar market for hedging risky payoffs resulting from uncertain demand should not be entirely surprising. The contracts observed in some of the industries studied by us further confirm the insight provided by our analysis. It is also logical that such contracts are seen for products that have short life cycle or are perishable such as grocery, personal computers and apparel, as these are the industries that are the most exposed to demand uncertainty. (The use of the single period inventory model as the decision making framework embodied in the newsperson problem is appropriate for such products as well.)

The rest of the chapter is organized as follows. We first provide a discussion on the relevant literature in the end of this section. In Sect. 9.2, we discuss the assumptions of our modeling framework and present the risk reducing contract for a single retailer (Sect. 9.2.1). In Sect. 9.3, we focus on the discrete setting in which there are a finite number of retailers that differ in their risk aversion magnitude and derive structural properties of the optimal menu of contracts. Section 9.4 presents results on how the population of retailers will be segmented on the basis of risk aversion. In Sect. 9.5, we switch to the continuous setting in which the retailers' risk aversion magnitude is drawn from a continuous distribution. This allows us to show that the menu proposed in Sect. 9.3 achieves the optimality in nearly all menus of contracts. Section 9.6 concludes our work and discusses some possible directions for future research.

9.1.1 Relevant Literature

In this section we review the relevant literature, including supply chain coordination and the pricing contracts.

9.1.1.1 Supply Chain Coordination

Recent research on contract mechanisms for achieving coordination in supply chains includes the work of Barnes-Schuster and Bassok (1996), Bassok and Anupindi (1997), Donohue (1996), and Moses and Seshadri (2000). Moses and Seshadri study the problem of coordinating the review period and order quantity in a supply chain consisting of a single manufacturer supplying many retailers from a distribution center. They show that coordination between the manufacturer and the retailers, on both the order quantity as well as the review period, can be achieved using credit terms alone as the policy variable. Donohue models a supply chain for fashion goods. There are two modes of supply, and a single supplier and a single buyer in the model. First, the buyer commits to an order quantity. Second, as more information becomes available, the buyer is allowed to change the order quantity on payment of a penalty for the deviation. Donohue shows that coordination can be achieved in this situation through pricing alone. Barnes-Schuster and Bassok too consider a supply chain with a single supplier and buyer. In their model, the buyer makes cumulative commitments to purchase over a time horizon. The contract includes a termination option. Similar price-quantity commitment models are studied by Bassok and Anupindi.

In contrast to these papers, the inclusion of risk aversion in supply chain coordination, and optimal design within a class of contracts are the novel features of our work. Moreover, our research is aimed at understanding how contracts, and in particular risk sharing plays a role in determining the distribution channel structure.

9.1.1.2 Pricing Contracts

Quantity discounts form an important pricing mechanism. A review of the marketing and the operations management literature reveals that there are four main reasons why quantity discounts are offered, namely: (i) to discriminate between retailers of different sizes or retailers with different holding costs (Oren et al., 1983), (ii) to reduce the transaction costs of the seller by inducing a larger order quantity from the buyer (Monahan, 1984; Lee and Rosenblatt, 1986; Lal and Staelin, 1984), (iii) to coordinate buyer seller transactions across single as well as multiple products, (Kohli and Park, 1989, 1994) and (iv) to mitigate loss in efficiency due to double marginalization and principal agent effects (Dolan, 1987; Jeuland and Shugan, 1983; Lal and Staelin, 1984; Porteus and Whang, 1991; Weng, 1995).

Traditional quantity discounts while widely used can not achieve complete coordination between the buyers and the seller for two reasons: (i) the (risk sharing) benefits as described by us can not be attained by simply varying the price as a function of the order quantity; and (ii) franchise fees are often required in addition to quantity discounts for achieving maximal channel profits (Oren et al., 1983; Weng, 1995). It is now well understood that such failures in coordination can arise as a result of using improperly constructed prices schedules. The theory of non-linear pricing (Wilson, 1993) can be brought to bear upon the problem is such situations. Quantity discounts and franchise fees are also examples of non-linear pricing. The menu of contracts necessary for inducing the retailers' participation in the distributor's network is yet another example of using a non-linear price schedule to separate out retailers with different attitudes towards risk.

9.2 Model, Assumptions, and Analysis

We consider the classical single period inventory problem. In this problem, the distribution of the demand as well as the retail price, p, are given. The retailer's problem is to choose the order quantity, S. In the contract, items are supplied at an initial unit price of c. If the demand in the period exceeds the quantity ordered, S, then the retailer obtains emergency shipments to cover any excess demand at a unit cost of e. On the other hand, if the demand is less than S, then the retailer sends the unsold items back to the seller, and obtains a credit of s per unit returned. The framework has been extended by us in Agrawal and Seshadri (2000) to incorporate risk aversion as well as to model multiple retailers as follows.

Model Assumptions

1. The retailers are alike in terms of demand distribution and cost parameters, and differ only with regard to their aversion to risk.
2. The retailer demands are independent and identically distributed random variables. The retail price (p) and the distribution of demand are unaffected by the contracts offered to the retailers.
3. Every contract has to be offered to every retailer. The retailer in turn selects a contract from the menu offered. This condition prevents direct (illegal) price discrimination by the distributor.
4. The retailers are not resellers, but purchase only to satisfy their own demand.
5. The distributor is risk-neutral (the distributor can be the manufacturer itself or the least risk averse retailer).
6. We use utility functions for money. Until Sect. 9.3 we make no additional assumption about the utility functions of retailers, except that the utility functions are concave and non-decreasing in the amount of wealth. From Sect. 9.3 onwards, the small gambles framework is adopted in our analysis (Pratt, 1964). In this framework the coefficient of risk aversion is assumed to be unaffected by the outcomes.

7. Retailers are expected utility (EU) maximizers, where, EU = expected value (EV) – the risk premium.
8. We assume that the general form of contracts that are made available to the retailers are given by $C(F, c, s, e)$, where,
   $F =$ Fixed side payment to the retailer,
   $c =$ Regular purchase price/unit,
   $s =$ Salvage value/unit of the unsold retailer stock,
   $e =$ Emergency purchase price/unit,

   and, $p \geq e \geq c \geq s$.
9. If two contracts provide the same EU to a retailer, the retailer will choose the contract that has the larger fixed payment. If the distributor offers a contract that provides the same EU as the ONC to the retailer, then the retailer will choose the contract offered by the distributor.

These modeling assumptions hold if (i) the product by itself contributes a small portion to the retailer's wealth (for Assumption 6), and (ii) retailers serve their local markets, and these markets have little or no overlap (for Assumption 2). Assumption 9 is standard in the analysis of such problems, for example see the discussion on p. 588, Kreps (1990). The assumption allows us to work with weak inequalities in Sect. 9.3.

9.2.1 A Single Retailer

In this section we consider the case of a single risk-averse retailer, and show that a risk neutral intermediary has an incentive to offer a risk sharing contract to the retailer. This has the effect of raising the ordering quantity to the EV maximizing value. The single period demand will be denoted as D, its distribution, mean and standard deviation as $F_D(\cdot)$, μ, and σ. The retailer's utility function for payoffs is $U(\cdot)$, a concave non-decreasing function. Let $\mathrm{E}[\cdot]$ stand for the expected value, and $[A]^+$ for the positive part of A. The random pay-off given the ordering quantity S, can be written as

$$\prod(S, F, c, s, e) = F + pD - cS + s[S - D]^+ - e[D - S]^+ .$$

The EU and EV maximizing order quantities, denoted as S_{opt}^U and $S_{\text{opt}}^{\text{EV}}$, are given by

$$\begin{aligned} S_{\text{opt}}^U(F, c, s, e) &= \arg\max_S \mathrm{EU}(pD - cS + s[S - D]^+ - e[D - S]^+) \\ S_{\text{opt}}^{\text{EV}}(F, c, s, e) &= \arg\max_S \mathrm{E}\left[pD - cS + s[S - D]^+ - e[D - S]^+\right] \\ &= F_D^{-1}\left(\frac{e - c}{e - s}\right). \end{aligned}$$

It is well known that (i) $S^{U}_{\text{opt}}(0,c,s,e) \leq S^{\text{EV}}_{\text{opt}}(0,c,s,e)$, and (ii) $E[\prod(S^{U}_{\text{opt}}, 0,c,s,e)] \leq E[\prod(S^{\text{EV}}_{\text{opt}}, 0,c,s,e)]$. The latter fact can be exploited by a risk neutral intermediary to act as a distributor as follows. Assume that the distributor takes the ONC, and in turn offers the contract $C(F,c',c',c')$, where, $p \geq c' \geq e$. In this contract the distributor offers to pay a fixed fee F to the retailer, and in addition charges a unit price c' for every unit sold. Under this arrangement the distributor decides the order quantity, S, and bears the costs of emergency shipment and of salvage.

Lemma 1 gives the conditions under which a class of contracts of the form $C(F,c',c',c')$ will be accepted by the retailer, and Lemma 3 shows that the distributor will use the EV maximizing value for the order quantity S.

Lemma 1. *Contract $C(F,c',c',c')$ is preferred by the retailer to the contract $C(0,c,s,e)$ if*

(i) $\mathrm{E}\left[\prod(S^{\text{EV}}_{\text{opt}}, F, c', c', c')\right] \geq \mathrm{E}\left[\prod(S^{U}_{\text{opt}}, 0, c, s, e)\right]$, *and*

(ii) $p \geq c' \geq e \geq s.$

This result can be strengthened as follows. Define "v to be more risk averse than u" when $v(\cdot) = k(u(\cdot))$, where $k(\cdot)$ is an increasing concave function, see for example Arrow (1974).

Theorem 1. *Given β and a retailer "x" who has an increasing utility function, $u(\cdot)$, such that this retailer is indifferent between the contracts $C(F-\beta,c',c',c')$ and the ONC, then every retailer with an increasing utility function $v(\cdot)$, who is more risk averse than retailer "x" will prefer $C(F-\beta,c',c',c')$ to the ONC.*

Given that the distributor can induce the retailer to take a "risk sharing" contract, the order quantity has to be decided upon by the distributor. We now prove that the optimal order quantity that the distributor will stipulate to the retailer is the EV maximizing solution to the original newsvendor problem.

Lemma 2. *The optimal ordering quantity for the distributor is $S^{\text{EV}}_{\text{opt}}(0,c,s,e)$.*

9.2.1.1 Multiple Retailers

In this section we consider the case when there are multiple retailers. Assume that the distributor offers the same contract $C(F,c',c',c')$ to all the retailers. We restrict attention to practical contracts in which $c' \leq p$. This class of contracts will be called C_{eq}. The contract $C(F,p,p,p)$ will be called a "risk free" contract because the retailer is paid a fixed fee F regardless of the quantity sold.

Theorem 2. *If the distributor offers a single contract to the whole population of retailers, and there are no diseconomies of scale in distribution, then the contract offered will be a risk free contract of the form $C(F,p,p,p)$. Moreover the contract that maximizes the distributor's expected profit need not be selected by the entire population of retailers.*

Remark: Risk pooling in supply chains has been studied by several researchers, see for example Eppen and Schrage (1981), Federgruen and Zipkin (1984), Jackson and Muckstadt (1989), Schwartz et al. (1985), and Schwartz (1989). As shown by their work, risk pooling leads to economies of scale in distribution under fairly general conditions. Therefore, from the viewpoint of minimizing the total cost of holding inventory, cost of emergency shipment and salvage cost, the distributor prefers to *add* more retailers to the distribution network. On the other hand, in order to attract more retailers, the distributor has to make increasingly attractive offers to *all* the retailers. Thus after attracting several of the most risk averse retailers to take a contract, the marginal profit to the distributor from inducing an extra retailer to accept the contract can become negative. The risk pooling and risk sharing effects therefore work in opposite directions.

We have carried out numerical investigations to understand just how the risk pooling and risk sharing effects interact when the distributor is constrained to offer a single contract of the form $C(F, c', c', c')$ to all the retailers. In these investigations, the demand is assumed to be normally distributed. To illustrate our point we consider complete risk pooling, i. e., the distributor can instantaneously and costlessly (other than the original costs of s, c, e), pool the inventory of all the retailers. The retailers are assumed to have CARA (constant absolute risk aversion) risk preferences and posses different degrees of risk aversion. The (retailers') coefficient of risk aversion is assumed to be uniformly distributed over a compact interval. The base case has $p = 11$, $e = 10$, $c = 9$, $s = 1$. The demand is normally distributed with mean $= 100$ and standard deviation $= 5$.

We summarize the results below and the details are shown in the attached figures. We find that as the demand becomes more volatile (i. e., σ/μ increases), Figs. 9.1 and 9.2, the distributor covers more retailers whereas risk-averse retailers order less when only the ONC is available. (The distributor's profit is shown if she covers only the retailers that are more risk-averse than the value on the x-axis.) These two (profit and order quantity under the ONC) are connected. The distributor's role becomes more important when there is greater loss of efficiency. The distributor is able to offer a better "deal" at a lower price, thus the marginal benefit of adding a less risk-averse retailer to the ones that are already covered by the risk-free contract is higher.

A similar effect is observed as the emergency cost e increases in Figs. 9.3 and 9.4. However, as the retailers' margin $p - c$ increases two effects work in opposite directions (see Figs. 9.5 and 9.6). As the profitability is higher, risk-averse retailers order less when price increases to obtain a "sure thing". That is, if the price is sufficiently high they will order nothing and wait to see the demand. Then, they order the exact quantity using the emergency shipment mode. Thus, initially as price increases more are covered by the distributor. But, once many retailers start ordering nothing the distributor prefers to cover only those and not the ones that order something due to the sharp difference in the compensation required to entice the additional ones to switch to the distributor.

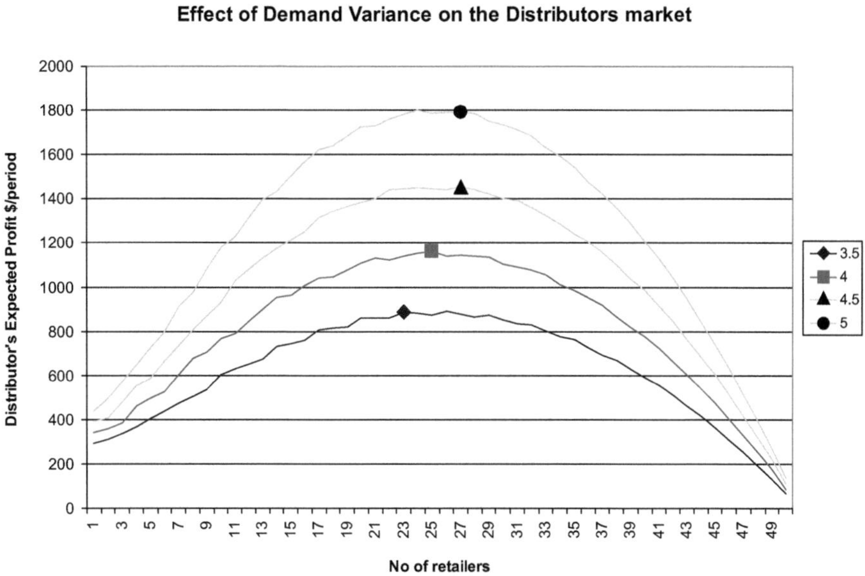

Fig. 9.1 Effect of volatility of demand on distributor's share of the market when a single contract is offered to all retailers

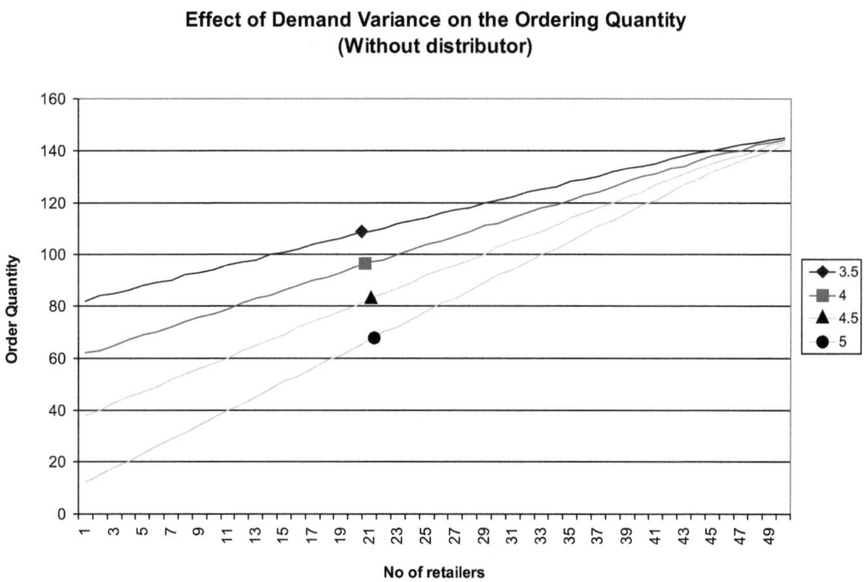

Fig. 9.2 Effect of volatility of demand on the order quantity in the absence of a distributor

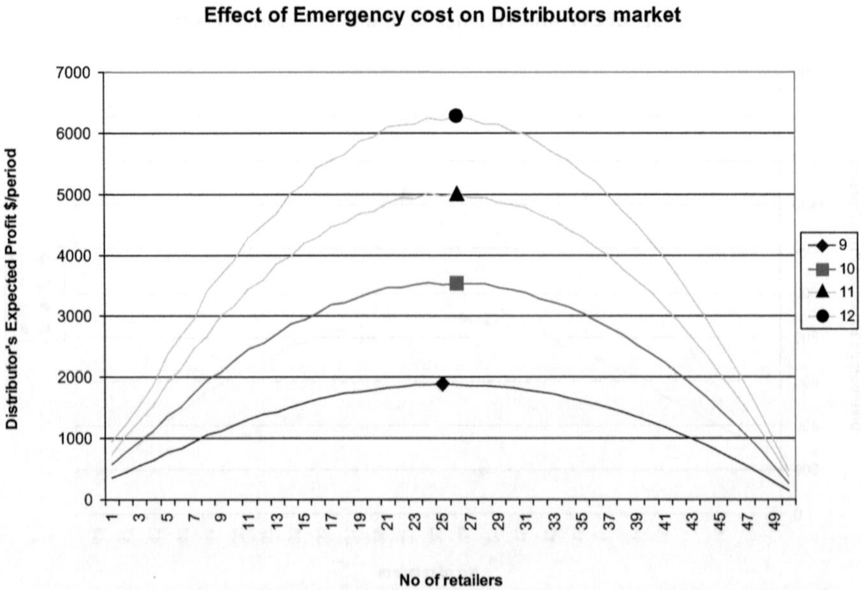

Fig. 9.3 Effect of emergency cost on distributor's share of the market when a single contract is offered to all retailers

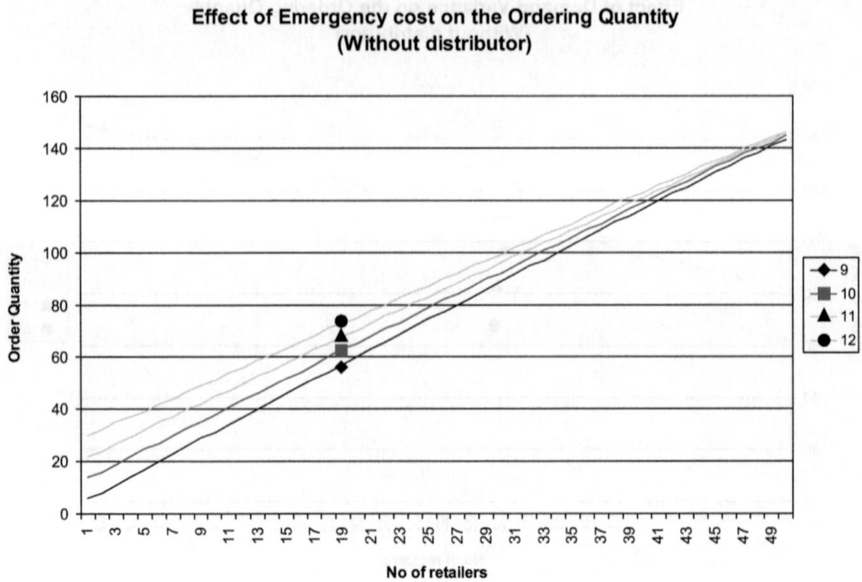

Fig. 9.4 Effect of emergency cost on the order quantity in the absence of a distributor

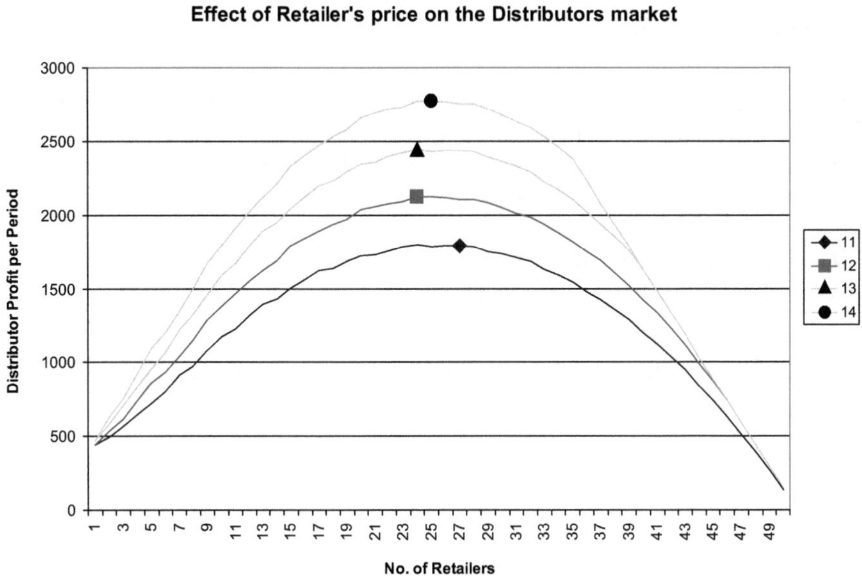

Fig. 9.5 Effect of selling price on distributor's share of the market when a single contract is offered to all retailers

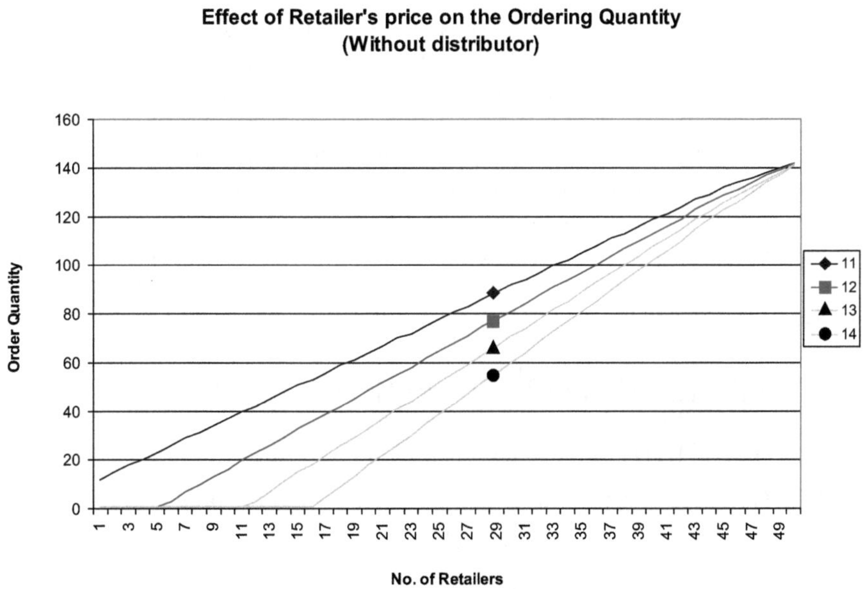

Fig. 9.6 Effect of selling price on the order quantity in the absence of a distributor

9.3 Multiple Contracts

The question that naturally follows Sect. 9.2 is whether there is an incentive (at all) for the distributor to offer contracts to all the retailers. In order to answer this question, we adopt the small gambles framework. We first consider the case in which there are a finite number of retailers that differ in their risk aversion magnitude. Later, in Sect. 9.5, we switch to the continuous case.

Consider a set of n retailers, denoted by N. The retailers are indexed by i and ordered in the decreasing order of risk aversion. Let

ρ_i: co-efficient of risk aversion; where $\rho_i \geq \rho_{i+1}$.
r_i: reservation utility for retailer i, which is defined as the expected utility derived by retailer i under the ONC, $C(0, c, s, e)$.

We assume that every retailer in N has a strictly positive order quantity under the ONC, and therefore has a strictly positive reservation utility. We shall consider contracts from the class C_{eq}. To simplify the notation, we denote a contract from this class as $C(F, c')$. The expected utility to retailer i from the contract $C(F, c')$ is given by

$$\mathrm{E}[U(\prod(S, F, c', c', c'))] = F + (p - c')\mu - \rho_i (p - c')^2 \sigma^2 / 2 .$$

The distributor offers the same menu of contracts to all the retailers in N. The menu will be written as $Q = \{(F_i, c_i)\}$. The set of retailers who accept a contract from this set, Q, will be denoted as $M(Q)$. To keep the notation simple, we also denote the contract *accepted* by retailer i from the set Q as (F_i, c_i). To focus on the risk sharing role played by the distributor, we have chosen not to model any scale economies obtained from risk pooling or from transportation. Therefore we assume that the distributor does not carry any inventory and trans-shipments between retailers are not allowed. There is no loss of generality in making this assumption as long as there are no scale diseconomies in distribution. The distributor's profit maximization problem is shown below.

$$\mathbf{P}: \quad \max_{Q} \sum_{i \in M(Q)} \left(\mathrm{E}\left[\prod\left(S_{\text{opt}}^{\text{EV}}, 0, c, s, e\right)\right] - (F_i + (p - c_i)\mu) \right)$$

$$s.t. \quad \text{If} \quad i \in M(Q) \Rightarrow F_i + (p - c_i)\mu - \rho_i (p - c_i)^2 \sigma^2 / 2 \geq r_i$$

$$F_i + (p - c_i)\mu - \rho_i (p - c_i)^2 \sigma^2 / 2 \geq F_j + (p - c_j)\mu - \rho_i (p - c_j)^2 \sigma^2 / 2$$

$$\forall i \in M(Q), \quad j \in Q .$$

The distributor's objective is to offer a menu that will maximize her expected profit. The first set of constraints, defines the set $M(Q)$ – a retailer will accept some contract from the menu of the contracts, only if his expected utility from the contract is at least as great as his reservation utility. The second set of constraints requires that the retailer will pick that contract from the menu which gives him the highest EU. We assume that Q consists of undominated contracts, that is, there is no contract in the menu which is strictly preferred to another by all retailers. We need the following properties of the retailers' reservation utilities for characterizing the set Q.

Lemma 3. *The reservation utilities, r_i's, are non-decreasing in i, and if $\rho_i > \rho_{i+1}$ then $r_i < r_{i+1}$.*

Lemma 4. *(i) r is a convex function of ρ, i. e., for $\rho_{i-1} > \rho_i > \rho_{i+1}$,*

$$\frac{r_{i+1} - r_i}{\rho_i - \rho_{i+1}} \geq \frac{r_i - r_{i-1}}{\rho_{i-1} - \rho_i}.$$

(ii) If $e > c > s, \sigma > 0$, and the demand distribution is continuous, then r is a strictly convex function of ρ.

We now state that every retailer will select a contract from the menu in the optimal solution to the distributor's problem. After showing this result, the precise characterization of the optimal menu of contracts will be given later.

Theorem 3. *Every retailer will be included in the set $M(Q)$ in the optimal solution to problem **P**.*

So far we have argued that it is in the distributor's interest to offer a menu such that every retailer selects a contract from it. Now we shall investigate the structure of the optimal menu of contracts.

Lemma 5. In the optimal (and undominated) menu of contracts, there will be (exactly) one risk free contract.

It should be noted that the risk free contract is a *less* expensive contract for the distributor to offer because the distributor does not have to pay any risk premium. However, the distributor has to offer all retailers the same menu of contracts. To entice all retailers and maximize profits simultaneously she may perforce have to the offer "riskier" contracts, with ($c_i < p$). On the other hand, the distributor always has the option of designing the risk free contract to attract more than just the most risk averse retailer. Let k be the number of retailers that take the risk free contract in the optimal menu.

In the next section we will discuss the question of determining the value of k to maximize the distributor's profit. Before we get to that question, we develop the optimal structure of the menu of contracts offered to retailers, $k+1, k+2, \ldots, n$. To obtain this characterization we make an additional assumption, namely that the reservation utility, r_i is an increasing and strictly convex function of ρ_i. This assumption holds good when $e > c > s, \sigma > 0, S_i > 0$, and the demand distribution is continuous.

Theorem 4. *For a given value of k (i. e., retailers $1, 2, ..., k$ accept the risk free contract, $F_k = r_k, c_k = p$), the distributor's profit is maximized by offering the* *contract* $(F_i, c_i), i \geq k+1$ *given by,*

$$c_i = p - \left(\frac{2(r_i - r_{i-1})}{(\rho_{i-1} - \rho_i)\sigma^2}\right)^{0.5},$$
$$F_i + (p - c_i)\mu - \rho_i (p - c_i)^2 \sigma^2 / 2 = r_i \,.$$

We summarize the properties of the optimal menu of contracts.

1. The distributor makes a profit from all contracts.
2. The prices charged to the retailers are decreasing in i, i.e., $c_i > c_{i+1}, i \geq (k+1)$.
3. The fixed side payments made to the retailers are decreasing in i, i.e., $F_i > F_{i+1}, i \geq (k+1)$.
4. From the fact that retailer i obtains the same EU from contracts (F_i, c_i) and (F_{i+1}, c_{i+1}), the EV's of the contracts (F_i, c_i) are increasing in $i \geq k$.
5. Retailers $1, 2, ..., k-1$ obtain EU's greater than their reservation utility. All other retailers get exactly their reservation utility.

9.4 Risk Aversion and Channel Structure

In this section we will discuss how many retailers should get the fixed contract, i.e., the decision variable is now k. The distributor's profit maximization problem is (see problem P):

$$\max_k \left(\begin{array}{l} \left(\mathrm{E}\left[\prod\left(S_{\mathrm{opt}}^{\mathrm{EV}}, 0, c, s, e\right)\right] - r_k\right) k \\ + \sum\limits_{i>k} \left(\mathrm{E}\left[\prod\left(S_{\mathrm{opt}}^{\mathrm{EV}}, 0, c, s, e\right)\right] - \left(r_i + \rho_i (p - c_i)^2 \sigma^2 / 2\right)\right) \end{array} \right),$$

where F_i and c_i are as defined in Sect. 9.3. This problem can be solved numerically using a search technique. However, further insight can be obtained by assuming that the coefficient of risk aversion can take values on the interval [0, 1], and has the density function $f_r(\rho)$. In this model, the fraction of retailers in the population whose coefficient of risk aversion lies in the interval, $[\rho, \rho + \mathrm{d}(\rho)]$ is given by $f_r(\rho)\mathrm{d}(\rho)$. The distribution function of risk aversion and its complement are denoted by F_r and F_r^c. Assume that the reservation utility is a continuously differentiable (convex) function of ρ. Passing to the limit, the cost $c(x)$ charged to the retailer with a coefficient of risk aversion equal to x will be given by,

$$c(x) = p - \left(-\frac{2\mathrm{d}r(x)}{\mathrm{d}x}\frac{1}{\sigma^2}\right)^{0.5}.$$

The profit function can then be restated as,

$$\begin{aligned}\Xi(x) =& \left(\mathrm{EV}\left(S_{\mathrm{opt}}^{\mathrm{EV}}, c, s, e\right) - r(x)\right)F_{\mathrm{r}}^{\mathrm{c}}(x) \\ &+ \int_x^0 \mathrm{EV}\left(S_{\mathrm{opt}}^{\mathrm{EV}}, c, s, e\right) - r(y) + y\frac{\mathrm{d}r(y)}{\mathrm{d}y}) f_{\mathrm{r}}(y)\mathrm{d}y\,. \\ \frac{\mathrm{d}\Xi(x)}{\mathrm{d}x} =& -\frac{\mathrm{d}r(x)}{\mathrm{d}x}\left(F_{\mathrm{r}}^{\mathrm{c}}(x) - x f_{\mathrm{r}}(x)\right).\end{aligned}$$

From the fact that $-\frac{\mathrm{d}r(x)}{\mathrm{d}x} > 0$, the maxima of the profit function $\Xi(x)$ are independent of the reservation utility. Moreover, if the function $xF_{\mathrm{r}}^{\mathrm{c}}(x)$ is uni-modal and has a unique maximum in the interior of [0, 1], then the optimal value of x is independent of the reservation utility.

Remark. The above assumption implies that the function $F_{\mathrm{r}}^{\mathrm{c}}(x) - xf_{\mathrm{r}}(x)$ is initially positive and then becomes and stays negative. Note that if we interpret x as the price and the complementary cdf as the effective demand, $xF_{\mathrm{r}}^{\mathrm{c}}(x)$ represents the revenue as a function of price. Its unimodality is commonly assumed in many papers on revenue management, and a sufficient condition for unimodality is when the distribution has the increasing generalized failure rate property (IGFR), namely, $xf_{\mathrm{r}}(x)/(1 - F_{\mathrm{r}}(x))$ is increasing in x. Ziya et al. (2004) compare three conditions that induce revenue unimodality, and mention some common distributions that satisfy these conditions such as normal, uniform, and gamma. The reader is referred to these papers and references therein for more details.

Remark. Note that this assumption is scale invariant, i. e., if x is scaled to bx then $xF_{\mathrm{r}}^{\mathrm{c}}(x/b)$ remains unimodal. Moreover, the 'point' that achieves the maximum is also scale invariant.

In particular, the above condition holds when ρ has a beta distribution with parameters (p, q). We therefore see that for a wide variety of unimodal and bi-modal distributions the optimal fraction of retailer population that select the risk free contract is independent of the product characteristics such as selling price or purchase costs. The optimal fraction for a few distributions is given below in Table 9.1.

Table 9.1 Optimal fraction of retailers who are given the risk free contract

Distribution of $f_{\mathrm{r}}(x)$	Optimal Value of x
Uniform (U[0,1])	$\frac{1}{2}$
Triangular with $f_{\mathrm{r}}(0) = 0$, $f_{\mathrm{r}}(1) = 2$	$\sqrt{\frac{1}{3}} = 0.577$
Triangular with $f_{\mathrm{r}}(0) = 2$, $f_{\mathrm{r}}(1) = 0$	$\frac{1}{3}$
Triangular and symmetric $f_{\mathrm{r}}(0) = f_{\mathrm{r}}(1) = 0$, $f_{\mathrm{r}}(0.5) = 2$	$\sqrt{\frac{1}{6}} = 0.408$
Truncated Normal with mean = 150, $\sigma = 5$	0.5

We illustrate the results using the same example considered earlier, with $p = 11$, $e = 10$, $c = 9$, $s = 1$ and demand normally distributed with mean of 150 and standard deviation of 5 and also for $p = 14$ with the rest of the parameters remaining the same. In Figs. 9.7 and 9.8 we show the reservation price and the optimal fixed side payment to entice a retailer to source from the distributor. In Figs. 9.9 and 9.10 we show the optimal variable cost to charge a retailer that has a given coefficient of risk-aversion. The distributor's profits are also shown assuming that the all retailer's with coefficient of risk-aversion higher than the value on the x-axis are given the risk free contract. As predicted by our analytical results, the optimal fraction of retailers that are given the risk free contract is independent of the retail price (in this case the fraction is always a half)!

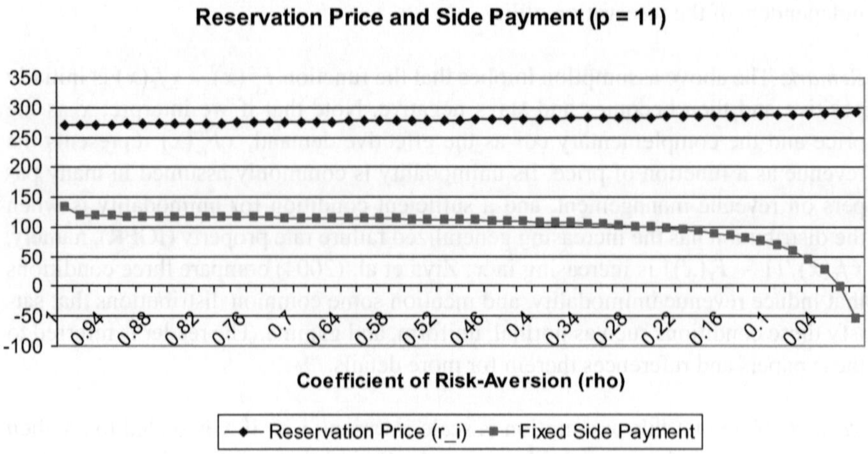

Fig. 9.7 Reservation price under the ONC and optimal side payment ($p = 11$)

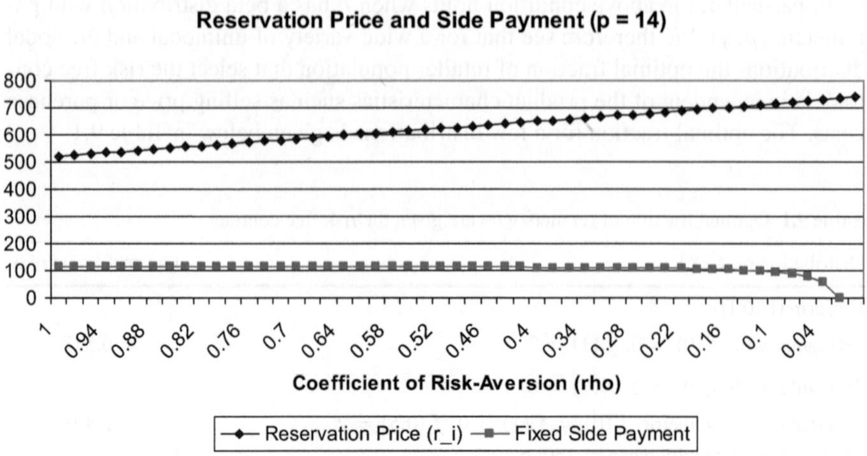

Fig. 9.8 Reservation price under the ONC and optimal side payment ($p = 14$)

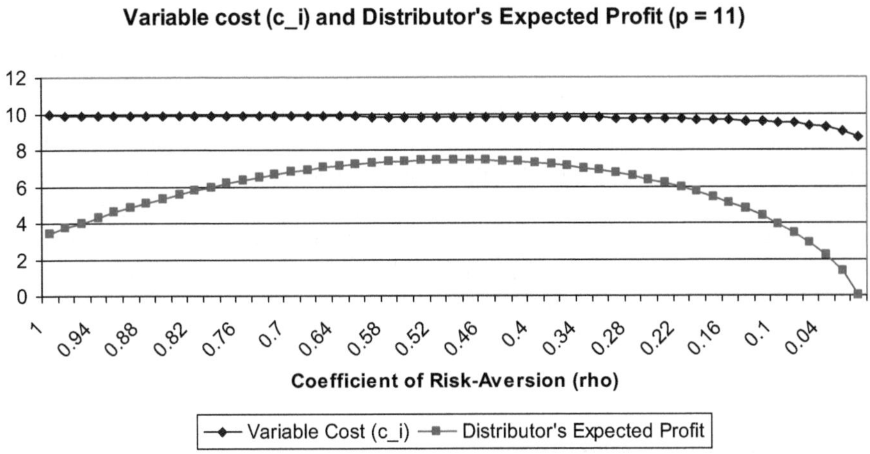

Fig. 9.9 Optimal c_i and distributor's expected profit as a function of fraction of retailers that obtain the risk-free contract ($p = 11$)

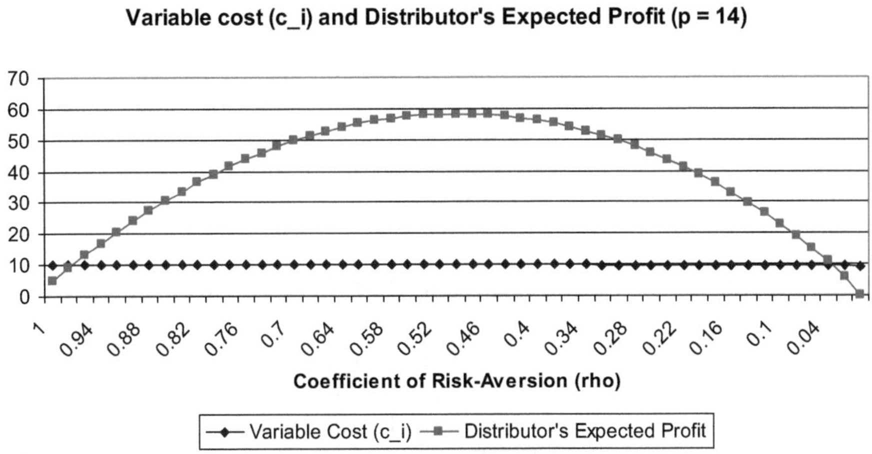

Fig. 9.10 Optimal c_i and distributor's expected profit as a function of fraction of retailers that obtain the risk-free contract ($p = 14$)

9.5 Continuous Formulation and the Optimality of the Menu

So far we have showed that under the proposed menu, less risky (from the demand risk perspective) contracts are given to more risk-averse retailers. Such a menu of contracts increases the distributor's expected profit because the retailers order more products. The intuitive content of this result is that the distributor can trade-off the expected value obtained by risk averse retailers against the gain in utility from risk reduction.

A natural question that arises is whether the menu of contracts is optimal. To this end, we consider an alternative setting in which the number of retailers is infinite and their coefficient of risk aversion is drawn from a continuous distribution. We label this scenario as the continuous formulation. Through this continuous formulation, we are able to apply optimal control theory to solve the contract design problem. In this section, we demonstrate that the optimal menu not only has the same structure as given above but is also optimal among nearly all contracts. We also show that the distribution of the risk aversion coefficient uniquely determines the channel structure. Thus, distribution systems for products with long supply lead times and short lifecycles should bear marked similarities reflecting the attitude towards risk of channel participants.

9.5.1 Continuous Type Space

In this section, we assume that the coefficient of risk aversion ρ can take values in the interval $[0, 1]$. We assume that it has the density function $f_{\mathrm{r}}(\rho)$. In this representation, the fraction of retailers in the population whose coefficient of risk aversion lies in the interval $[\rho, \rho + \mathrm{d}\rho]$ is given by $f_{\mathrm{r}}(\rho)\mathrm{d}\rho$. The distribution function of risk aversion and its complement are denoted by F_{r} and $F_{\mathrm{r}}^{\mathrm{c}}$. We also assume that the reservation utility is a differentiable and convex function of ρ. This continuous setting is adopted from Chen and Seshadri (2006).

9.5.2 Optimal Contract Menu in the Continuous Case

First, assume that *all* retailers are offered a contract and focus on the design of the optimal menu of contracts. We formulate the optimal contract design problem in two stages. In the first stage, we assume that there exists a constant $\tau \in [0, 1]$ such that retailers with $\rho \in [\tau, 1]$ will choose the risk-free contract $(r(\tau), p, p, p)$ from the menu, where $r(\rho)$ is the reservation utility of retailer with risk aversion coefficient ρ. Note that $(r(\tau), p, p, p)$ is the cheapest risk-free contract that satisfies the participation constraints for all retailers $\rho \in [\tau, 1]$ for two reasons: It is risk-free, so has the lowest expected value of all contracts that provide a utility of $r(\tau)$. We show below that it provides utility greater than or equal to $r(\rho)$ for all retailers with ρ in $[\tau, 1]$. We first develop optimal incentive compatible contracts to every retailer $\rho \in [0, \tau]$ under such assumptions. In the second stage, we optimize over the choice of τ. Notice that we do not exclude the possibility of $\tau = 1$, i.e., only the most risk-averse retailer is offered the risk-free contract, and hence this is without loss of generality.

Consider any menu such that $g(x)$ and $h(x)$ are respectively the mean and variance of the payoff to a retailer if menu item x is chosen, where x can take values in the interval $[0, \tau]$ and $\tau \in [0, 1]$. We therefore consider the most general form of the contracts. This is the most general form because retailers are concerned only about the mean and the variance of the payoff. With some abuse of notation, let a retailer with coefficient of risk aversion equal to $x \leq \tau$ choose menu item x. Given a fixed τ, the distributor's problem is to choose $\{(g(\rho), h(\rho)), \rho \in [0, \tau)\}$ that solves the following maximization problem:

$$\max\left\{ \left[\mathrm{EV}\left(S_{\mathrm{opt}}^{\mathrm{EV}}, c, s, e\right) - r(\tau)\right] F_{\mathrm{r}}^{\mathrm{c}}(\tau) + \int_0^\tau \left(\mathrm{EV}\left(S_{\mathrm{opt}}^{\mathrm{EV}}, c, s, e\right) - g(\rho)\right) f_{\mathrm{r}}(\rho)\mathrm{d}\rho \right\},$$

$$\begin{aligned}
\text{s.t. } & (\text{IC-1})\ \rho \in \operatorname{argmax}_{z\in[0,\tau)} g(z) - \rho h(z), \quad \forall \rho \in [0, \tau], \\
& (\text{IC-2}) r(\tau) \geq \max_{z\in[0,\tau)} g(z) - \rho h(z), \quad \forall \rho \in [\tau, 1], \\
& (\text{IR-1}) g(\rho) - \rho h(\rho) - r(\rho) \geq 0, \quad \forall \rho \in [0, \tau], \\
& (\text{IR-2}) r(\tau) \geq r(\rho), \quad \forall \rho \in [\tau, 1],
\end{aligned}$$

where $S_{\mathrm{opt}}^{\mathrm{EV}} \equiv F_D(\frac{e-c}{e-s})$ ($F_D(\cdot)$ denotes the demand distribution) is the expected value maximizing order quantity. $\mathrm{EV}(S_{\mathrm{opt}}^{\mathrm{EV}}, c, s, e)$ is the expected cost that the distributor has to pay for buying the vendors' ONC.

In the above equation, the first two inequalities are incentive compatibility (IC) conditions for respectively the retailers that receive a specific contract designed for her and the retailers that accept the risk-free contract. In (IC-1), we say that the contract menu is incentive compatible since the utility of retailer ρ is maximized if she chooses the contract with mean $g(\rho)$ and variance $h(\rho)$. On the other hand, $r(\tau)$ is the utility of retailer $\rho \geq \tau$ when she receives the risk-free contract, and (IC-2) guarantees that she prefers this to any other contracts $g(z), h(z)$ with $z \in [0, \tau]$.

The last two inequalities represent individual rationality (IR) conditions, i. e., each retailer shall get at least her reservation utility. Note that the reservation utility $r(\rho)$ can be explicitly expressed as

$$r(\rho) = \max_S \left\{\mathrm{E}\left[\prod(S, 0, c, s, e)\right] - \rho\frac{\mathrm{Var}[\prod(S,0,c,s,e)]}{2}\right\},$$

where $\prod(S, 0, c, s, e)$ is the profit if the ONC is accepted and the order quantity is S. It can be verified that $r(\rho)$ is strictly decreasing in ρ, and hence the last inequality (IR-2) is automatically satisfied.

The following theorem summarizes our results thus far.

Theorem 5. *Retailers with $\rho \in [\tau, 1]$ choose contract $(r(\tau), p, p, p)$, where constant $\tau \in [0, 1]$, and the distributor has to serve all retailers. Then the necessary conditions for the optimal contract menu are (i) (LO), and (ii) retailers $\rho \in [0, \tau]$ receive their reservation utilities.*

9.5.2.1 Candidate Menu

Now we will propose a candidate menu of contracts. The inspiration is due to the optimal menu in the discrete version, i. e., the one proposed in Theorem 4.1.2. We will focus on the class of contracts with a fixed franchise fee and common cost $\{F(\rho), c(\rho)\}$, and prove that this class is broad enough to achieve the optimality. The cost $c(\rho)$ charged to the retailer with a coefficient of risk aversion equal to ρ and the corresponding fixed side payment $F(\rho)$ are given by the solution to

$$c(\rho) = p - \left(-\frac{2\mathrm{d}r(\rho)}{\mathrm{d}\rho}\frac{1}{\sigma^2}\right)^{0.5},$$
$$F(\rho) + (p - c(\rho))\mu - \rho\frac{(p - c(\rho))^2\sigma^2}{2} = r(\rho)\,. \tag{9.1}$$

The corresponding $g(\rho)$ and $h(\rho)$ are $F(\rho) + (p - c(\rho))\mu$ and $\frac{(p-c(\rho))^2\sigma^2}{2}$, $\forall \rho \in [0, \tau)$.

It can then be verified that the proposed contract menu satisfies the necessary and sufficient conditions. Note that the continuous formulation allows us to apply the optimal control theory and consequently the verification is greatly simplified as opposed to the discrete formulation.

9.5.2.2 Optimal Choice of τ

Now we turn to the second stage: optimizing over the choice of τ. Let $\Xi(\tau)$ denote the profit function of the distributor when retailers that have a coefficient of risk aversion greater than τ are offered the risk-free contract. Recall that $\Xi(\tau)$ can be restated as

$$\Xi(\tau) = \left(\mathrm{EV}\left(S_{\mathrm{opt}}^{\mathrm{EV}}, c, s, e\right) - r(\tau)\right) F_{\mathrm{r}}^{\mathrm{c}}(\tau)$$
$$+ \int_{\tau}^{0} \left(\mathrm{EV}\left(S_{\mathrm{opt}}^{\mathrm{EV}}, c, s, e\right) - r(\rho) + r\frac{\mathrm{d}r(\rho)}{\mathrm{d}\rho}\right) f_{\mathrm{r}}(\rho)\mathrm{d}\rho\,.$$

Using the rule for differentiating under the integral we obtain

$$\frac{d\Xi(\tau)}{d\tau} = -\frac{\mathrm{d}r(\tau)}{d\tau}\left(F_{\mathrm{r}}^{\mathrm{c}}(\tau) - \tau f_{\mathrm{r}}(\tau)\right)\,.$$

From the unimodality of $\tau F_{\mathrm{r}}^{\mathrm{c}}(\tau)$, the maxima of the profit function are independent of the reservation utility. In other words, the fraction of retailers who select the risk free contract is independent of *product characteristics* if the distribution is unimodal.

Recall that $k \in (0, 1)$ is the value of τ at which the function $\tau F_{\mathrm{r}}^{\mathrm{c}}(\tau)$ attains its maximum. Thus,

$$F_{\mathrm{r}}^{\mathrm{c}}(\tau) - \tau f_{\mathrm{r}}(\tau) \geq 0\,, \quad \tau \in [0, k]\,,$$

and the necessary condition for optimality is

$$\tau = k\,.$$

We use $C^* = \{F^*(\rho), c^*(\rho)\}$ to denote the contract menu. Notice that in the continuous case the menu C^* we propose again gives a risk-free contract to all retailers with coefficient higher than k. This completes the characterization of the optimal menu of contracts, and therefore we have

Theorem 6. *Let* $k = \text{argmax}_{\tau \in [0,1]} \tau F_{\text{r}}^{\text{c}}(\tau)$ *and* $C^* = \{(F^*(\rho), c^*(\rho)), \rho \in [0, k]\}$. *Then the proposed* C^* *is optimal among the class of menus that serve all retailers. Moreover, under the optimal menu of contracts, all retailers* $\rho \in [0, k]$ *receive their reservation utilities, and retailers* $\rho \in (k, 1]$ *are offered the same risk-free contract.*

Note that the class of menus we consider include all menus since retailers' utility functions are of the mean-variance format. Hence, if all retailers ought to be served, C^* is indeed the optimal menu.

9.5.2.3 Verification of Optimality

We have shown that if all retailers are served, our proposed contract menu C^* yields the highest expected payoff to the distributor. The purpose of this section is to show that our proposed menu of contracts is indeed optimal even when we allow the distributor to exclude some retailers (for example, offer contracts only to those whose coefficient of risk aversion falls in $[0, 0.25) \cup [0.7, 0.993)$). We do this through three lemmas and a theorem as stated below.

Let $S(C)$ be the set of retailers that receive and accept contracts from the menu C. For each $x \in S(C)$, the menu C specifies a bundle $(F(x), c(x))$. Needless to say, the sets $S(C)$ of interest should be measurable with respect to the probability space $([0, 1], B, F_{\text{r}}(\cdot))$. Due to the special structure of our proposed contract, we show that if the distributor wants to serve merely the retailers on an interval $I \subset [0, 1]$ and ignore all other retailers, the optimal one-segment contract menu coincides with the proposed contract C^* restricted to the interval I (denoted as $C^*|_I$):

Lemma 6. (DECOMPOSITION) *Suppose* $C^* = (F^*(x), c^*(x))$ *is the optimal contract menu for* $S(C^*) = [0, 1]$. *Then for any interval* $I \subset [0, 1], C^*|_I$ *is also optimal.*

This lemma says for any given contract C with arbitrary number of segments, the distributor will be better off if she replaces C by menu C^* in every segment. Next we will study two properties of the proposed contract menu C^*, namely the no-skip property and push-to-the-end property.

Lemma 7. (NO-SKIP PROPERTY) *Suppose the distributor adopts menu* C^* *and* $S(C^*)$ *is composed of two disjoint intervals* I_1 *and* I_2, *then the distributor will be*

better off by offering contracts to all retailers in I_1, I_2, and also those between I_1 and I_2.

Applying this lemma inductively, we obtain that if the distributor offers the menu C^*, then the optimal $S(C^*)$ will be an interval. The following lemma says that while offering family of contracts C^*, the distributor should not leave any uncovered intervals of retailers from both ends.

Lemma 8. (PUSH-TO-THE-END PROPERTY) *Suppose the distributor adopts menu C^* and $S(C^*)$ is nonempty. Let $\bar{s} \equiv \sup\{x : x \in S(C^*)\}$ Then it is in the distributor's interest to set $\bar{s} = 1$. On the other hand, if $\underline{s} \equiv \inf\{x : x \in S(C^*)\}$, the distributor will set $\underline{s} = 0$.*

Combining the above three lemmas, if the distributor offers $C^* = \{F^*(x), c^*(x)\}$, she will offer contracts to the entire interval [0, 1] to maximize her profit. Bearing in mind the structure of C^*, we can then claim that the proposed contracts $(F^*(x), c^*(x))$ are optimal among all contracts that offer a menu to a measurable set of retailers.

In our model, the reservation utility of a retailer comes from her alternative "accepting the ONC". Therefore, the reservation utility varies from type to type in nature, and is decreasing in ρ. The optimal contract menu C^* enables the distributor to extract all the information rent of retailers who are less risk averse, while leaving the retailers with higher risk aversion the full information rent. This result is in strict contrast to the standard case in the nonlinear pricing literature where players are endowed with the same reservation utilities. If the reservation utility is the same, the theory predicts that the most risk-averse retailer receives just the reservation utility, and everybody else with $\rho < 1$ enjoys the information rent. Since the reservation utility is decreasing in ρ, the optimal contract has to match the participation constraint for retailers with lower ρ to induce their participation, and distort the contract terms for retailers with higher ρ. By doing so the distributor incurs a lower cost and maximizes her profit.

The fact that a continuum of retailers receive a risk-free contract is also worth noting. It is known as the "bunching" phenomenon, which may occur in the standard case when the monotone hazard rate property of types fails. Here the bunching occurs in retailers with high risk aversion and the contract offers them the efficient level, i. e., it fully covers the demand risk for those risk averse retailers.

Finally, it can be shown that the proposed menu C^* is unique up to a measure-zero modification, which means all menus properly different from C^* are suboptimal.

9.6 Future Research and Conclusion

This chapter demonstrates that an important role of an intermediary in distribution channels is to reduce the risk faced by retailers. The sharing of risk can be achieved by offering mutually beneficial risk sharing contracts which also raise the retailers' order quantity to the expected value maximizing quantity. Thus inefficiency created due to risk aversion on part of the retailers can be avoided. Through our continuous framework, we show that the contract menu is optimal among all possible menus, provided that the distribution of risk aversion is continuous and satisfies some mild condition commonly adopted in the revenue management literature.

We have shown that it is to the benefit of the distributor to offer a completely risk free contract to (one or more) of the most risk averse retailers. In practice this can be interpreted as an integrated channel in which the distributor owns the retail channel. For a variety of unimodal and bimodal distributions, the fraction of the retailer population who are offered the risk free contract is dependent only on the distribution of the coefficient of risk aversion of the retailers; and not dependent on the product parameters such as costs or revenue. This implies that in these cases, the retailer population can be segmented on the basis of the retailers' risk profile.

We have also demonstrated an important decentralization result. The distributor is responsible for the ordering decision in our model, and the retailer is shown to be content with this arrangement. Therefore, even if there are economies of scale in distribution, the distributor will offer the menu of contracts, and independently optimize with respect to her distribution costs.

There are still many open questions in this area. Of particular interest is the case when there is competition between the distributors. When multiple distributors compete for the position of risk intermediation, offering the menu of contracts proposed in this chapter is no longer the best strategy for each distributor. This is because some retailers earn just their reservation utilities and therefore a distributor may be willing to offer a better contract that not only leaves some surplus for the retailers but also gives rise to a positive expected payoff for the distributor. In doing so, the retailers will accept this new contract instead and the distributor benefits from capturing the business. The above argument shows that the competition shall significantly alter the equilibrium contract offers. More importantly, as discussed in Rothschild and Stiglitz (1976) there may not even be an equilibrium in this market.

Another case of interest is when the price is a decision variable (i. e., the retailers are price setters). This is a very practical scenario since in many industries, retailers have certain local monopoly power and therefore are able to set the retail prices at their own discretion. In this case, the presence of the distributor should affect the retailers' pricing decisions. In particular, when the distributor attempts to capture certain portion of the system profit, the profit margin seen by the retailers becomes smaller compared to the case without the distributor. In this case, we expect that the presence of the distributor may result in higher retail prices and lower consumption. A related issue is what if the retailers are able to exert costly effort to enhance the demand. How to induce appropriate efforts by designing contract menus is a critical issue for the distributor is such a scenario.

Finally, in our framework, the retailers are homogeneous and independent except that they have different risk aversion magnitude. This is definitely a simplified assumption that allows us to derive the analytical results. In reality, the supply chain may consist of retailers that face different sizes of markets, and the demands across different markets may even be correlated. These factors may also affect the supply chain structure and the optimal contracts between the distributor and the retailers.

References

Agrawal, V., Seshadri, S. (2000) Risk intermediation in supply chains. *IIE Transactions* 32:819–831

Andel, T. (1996) Manage Inventory, Own Information. *Transportation and Distribution* 37(5):54–56

Arrow, K. J. (1974) *Essays in the Theory of Risk Bearing*, North-Holland, New York, NY

Baron, D. P. (1973) Point Estimation and Risk Preferences. *J. the American Statistical Association* 68:944–950

Bassok, Y., Anupindi, R. (1997) Analysis of Supply Contracts with Total Minimum Commitment. *IIE Transactions* 29(5):373–381

Barnes-Schuster, D., Bassok. Y. (1996) Single Product Supply Contracts with Total Minimum Commitment and Option to Terminate. *MSOM Conference Proceedings*, Dartmouth College, 282–287

Bickel, P., Lehmann, E. (1979) Descriptive Statistics for Non-Parametric Models. IV. Spread," in *Contributions to Statistics*, Jane Juneckova (Ed.), Reidel Publishing Co.

Bird, L., Bounds, W. (1997) Stores' Demands Squeeze Apparel Companies. *Wall Street Journal*, Eastern edition, Jul 15, B1:3

Carvajal, D. (1998) Publishing Industry's Focus Shifts to Limiting Returns. *New York Times* Late Edition (East Coast) Jul 27, D9:1

Chen, Y.-J. Seshadri, S. (2006) Supply chain structure and demand risk. *Automatica* 42(8): 1291–1299

Dolan, R. J. (1987) Quantity discounts: Managerial issues and research opportunities. *Marketing Science*, 30(6):1–22

Donohue, K. (1996) A Two Stage Model Of Supplier-Buyer Coordination Under Forecast Uncertainty. *MSOM Conference Proceedings*, Dartmouth College, 20–25

Eeckhoudt, L., Gollier, C., Schlesinger, H. (1995) The Risk Averse (and Prudent) Newsboy. *Management Science*, 41(5):786–794

Eppen, G. D., Schrage, L. (1981) Effects of Centralization in a Multi-warehouse System with Lead Times and Random Demand. in Schwarz, L. B., *Multi-Level Inventory/Production Control Systems: Theory and Practice*, TIMS Studies in the Management Sciences, North-Holland, Amsterdam

Federgruen, A., Zipkin, P. (1984) Approximations of Dynamic Multilocation Production and Inventory Problems. *Management Science* 69–84

Horowitz, I. (1970) *Decision Making and the Theory of the Firm*, Holt, Rinehart, and Winston, New York, NY

Jackson, P., Muckstadt, J. (1989) Risk Pooling in a Two-Period, Two-Echelon Inventory Stocking and Allocation Problem. *Naval Res. Logistics Quarterly* 36:1–26

Jeuland, A., Shugan, S. (1983) Managing Channel Profits. *Marketing Science* 2:239–272

Kohli, R., Park, H. (1994) Coordinating Buyer-seller Transactions Across Multiple Products. *Management Science* 40(9):1145–1150

Kohli, R., Park, H. (1989) A Cooperative Game Theory Model of Quantity Discounts. *Management Science* 35(6):693–707

Kreps, D. (1990). *A Course in Microeconomic Theory*, Princeton University Press, NJ

Kirkpatrick, D. (1997) Now everyone in PCs wants to be like Mike. *Fortune* 136(5):91–92

Lal, R., Staelin, R. (1984) An Approach For Developing An Optimal Discount Policy. *Management Science* 30(12):1524–1539

Landsberger, M., Meilijson, I. (1994) The generating process and an extension of Jewitt's location independent risk concept. *Management Science* 662–669

Lee, H., Rosenblatt, M. (1986) A Generalized quantity discount pricing model to increase supplier's profits. *Management Science* 32(9):1177–1185

Lucas, A. (1996) Shelf wars. *Sales and Marketing Management* 148:3

Monahan, J. (1984) A Quantity Discount Pricing Model to Increase Vendor Profits. *Management Science* 30:720–726

Moses, M., Seshadri, S. (2000) Policy parameters for supply chain coordination. *IIE Transactions and Operations Engineering* 32(3):245–262

Oren, S., Smith, S., Wilson, R. (1983) Competitive Nonlinear tariffs. *Journal of Economic Theory* 29:49–71

Porteus, E. L., Whang, S. (1991) On Manufacturing/Marketing Incentives. *Management Science* 37:1166–1181

Pratt, J. W. (1964) Risk aversion in the small and in the large. *Econometrica* 122–136

Rothschild, M., Stiglitz, J. (1976) Equilibrium in Competitive Insurance Markets: An Essay on the Economics of Imperfect Information. *Quarterly Journal of Economics* 80:629–649

Schwarz, L. B. (1989) A Model for Assessing the Value of Warehouse Risk-Pooling: Risk- Pooling over Outside Supplier Lead-times. *Management Science* 35:828–842

Schwarz, L. B., Deuermeyer, B., Badinelli, R. (1985) Fill-Rate Optimization in a One-Warehouse N-Identical Retailer Distribution System. *Management Science* 31:488–498

Shaked, M., Shanthikumar, J. (1994) *Stochastic Orders and Their Applications*, Academic Press, CA

Stiglitz, J. E. (1977) Monopoly, Non-linear Pricing and Imperfect Information: The insurance Market. *Review of Economic Studies* 44:407–30

Weng, Z. K. (1995) Channel Coordination and Quantity Discounts. *Management Science* 41(9):1509–1522

Wilson, R. (1993) *Non-Linear Pricing*, Oxford University Press, New York

Ziya, S., Ayhan, H., Foley, R. (2004) Relationships among three assumptions in revenue management. *Operations Research* 52:804–809

Chapter 10
Forecasting and Risk Analysis in Supply Chain Management: GARCH Proof of Concept

Shoumen Datta, Don P. Graham, Nikhil Sagar, Pat Doody, Reuben Slone, and Olli-Pekka Hilmola

This chapter is dedicated to Sir Clive W.J. Granger (September 4, 1934–May 27, 2009).

Abstract Forecasting is an underestimated field of research in supply chain management. Recently advanced methods are coming into use. Initial results presented in this chapter are encouraging, but may require changes in policies for collaboration and transparency. In this chapter we explore advanced forecasting tools for decision support in supply chain scenarios and provide preliminary simulation results from their impact on demand amplification. Preliminary results presented in this chapter, suggests that advanced methods may be useful to predict oscillated demand but their performance may be constrained by current structural and operating policies as well as limited availability of data. Improvements to reduce demand amplification, for example, may decrease the risk of out of stock but increase operating cost or risk of excess inventory.

Key words: Forecasting; SCM; demand amplification; risk management; intelligent decision systems; auto-id data; GARCH; RFID; operations management

10.1 Introduction

Uncertainty fuels the need for risk management although risk, if adequately measured, may be less than uncertainty, if measurable. Forecasting may be viewed as a bridge between uncertainty and risk, if a forecast peels away some degrees of uncertainty but on the other hand, for example, may increase the risk of inventory. Therefore, forecasting continues to present significant challenges. Boyle et al. (2008) presented findings from electronics industry, where original equipment manufacturers (OEM) could not predict demand beyond a 4 week horizon. Moon et al. (2000) presented demand forecasting from Lucent (Alcatel-Lucent), demonstrating improvement in forecasting accuracy (60 % to 80–85 %). Related observations (Datta, 2008a) resulted in inventory markdowns.

Availability of increasing volumes of data (Reyes et al., 2007; Reyes and Frazier 2008) demands tools that can extract value from data. Recent research has shown that advanced forecasting tools enable improvements in supply chain performance (Zhao et al., 2001; Zhao et al., 2002; Bayraktar et al., 2008; Wright and Yuan, 2008), if certain pre-requisites are optimized (ordering policies, inventory collaboration). Autoregressive models have been effective in macroeconomic inventory forecasts (Albertson and Aylen, 2003). Zhao et al. (2002) and Bayraktar et al. (2008) emphasize that the role of forecasting in supply chain is to indicate the right direction for the actors rather than being exactly right, at every moment. Choosing the correct forecasting method is often a complex issue (Chatfield and Yar, 1988).

The purpose of this work is to explore how advanced forecasting methods could be applied in global supply chain management and their requirements. We present real world results and use simulation of a 4-stage supply chain model, beer-game (Vensim simulation). We have also used SPSS statistical analysis software to construct autoregressive forecasting models. The problems may be described as: (1) how to construct autoregressive forecasting models for a supply chain environment and (2) what changes may be needed in supply chain design to apply these advanced forecasting models? In the next section, we introduce a few challenges and Sect. 10.3 discusses demand amplification in supply chain management. In Sect. 10.4 we discuss features of autoregressive models and generalized autoregressive conditional heteroskedasticity (GARCH). Data and analysis from a supply chain inventory model using GARCH is presented and although the results are preliminary, they are encouraging. Concluding thoughts and further research issues are proposed in Sect. 10.5.

10.2 Supply Chain Management and Demand Amplification

Despite rapid advances in SCM and logistics, inefficiencies still persist and are reflected in related costs (Datta et al., 2004). In developing nations the actual amounts are lower, but proportional share is higher (Barros and Hilmola, 2007). One of the logistically unfriendly country groups are oil producers (Arvis et al., 2007).

The high cost for operations offer prosperity for the service providers. In 2006, AP Moller-Maersk raked in US$ 46.8 billion in revenues (Hilmola and Szekely 2008). Deutsche Post reported revenues of € 63.5 billion in 2007 (Annual Report, 2007). Profitability and growth of these services are increasing, fueled by globalization. Global transportation growth exceeds global GDP growth (United Nations, 2005, 2007), since trade grows twice as fast as GDP. For decades companies emphasized lower inventories and streamlined supply chains but it has resulted in a situation (Chen et al., 2005; Kros et al., 2006) where material handling in distribution centers has increased (transportation growth combined with lower lot sizes). Management science and practice continues to explore ways to decrease transaction costs (Coase 1937, 1960, 1972, 1992) through real-time information arbitrage. Cooper and Tracey (2005) reported that in the 1990's Wal-Mart had an information exchange capacity of 24 terabytes. While massive investments in information technology (IT) may be prudent, the sheer volume of data begs to ask the question whether we have the tools to separate data from noise and if we have systems that can transform data into decisionable information.

In supply chain management, the issue of demand amplification or Bullwhip effect has been in the forefront for some time (Forrester, 1958; Lee et al., 1997) but it took decades before its importance was recognized. The development of supply chain management (Oliver and Webber, 1982; Houlihan, 1985) catalysed by globalization, highlighted the strategic importance of logistics and pivotal role of information technology. Small demand changes in the consumer phase resulted in situations, where factories and other value chain partners faced sudden peaks and down turns in demand, inventory holdings and a corresponding impact on production and delivery (delivery structure phenomenon due to Bullwhip effect is referred to as "reverse amplification" by Holweg and Bicheno (2000) and Hines et al. (2000) referred to it as "splash back"). Human intervention to tame the Bullwhip effect, compared to simple heuristics, leads to higher demand amplification (Sterman 1989).

It follows that demand amplification may have serious consequences due to increased uncertainty and increases the significance of risk management. During a down turn, Towill (2005) showed that amplification causes possible shortages on product volume (products are not ordered, even if demand is undiminished) and variety as well as on idle capacity in operations and involves potential layoff costs. In the case of positive demand, Towill (2005) identifies that stock deterioration and sales cannibalization produces lost income.

Consumers purchase products lured by discounts and that diminishes sales in the following time periods or seasons (Warburton and Stratton, 2002). During upswings, operations cost a premium for manufacturing and distribution (orders increase rapidly), but also decreases productivity development and increases waste levels. Recent emphasis on outsourcing and large-scale utilization of low cost sourcing has worsened demand amplification (Lowson, 2001; Warburton and Stratton, 2002; Stratton and Warburton, 2003; Hilletofth and Hilmola, 2008). Risks associated with production and transportation delays are considerably higher. To mitigate such risks, some corporations are using responsiveness as a strategic differentiator and have built their supply chains to react on market changes through more localized supply networks, for example, Benetton (Dapiran, 1992), Zara (Fraiman and Singh, 2002), Griffin (Warburton and Stratton, 2002; Stratton and Warburton, 2003), Obermeyer (Fisher et al., 1994) and NEXT (Towill, 2005). The carbon footprint of sourcing strategies will become increasingly relevant in view of future legislation. Logistics may ultimately benefit from a disruptive innovation (Datta, 2008b) in energy sourcing and management using wireless sensors networks (Datta, 2008e).

In recent decades, even macroeconomists are including inventory as a key indicator of economic decline of national economies (Ramey, 1989; Albertson and Aylen, 2003). Ramey (1989) argued that manufacturing input inventories, raw materials and work in process (WIP), fluctuated most in recession, while end-item or finished goods inventories fluctuate less (Table 10.1). However, the labour market volatility is also an issue in changing economic environments (Ramey, 1989). Although,

Table 10.1 All numbers billions of US-dollars (1972), annual rates of change

Recessions		Retail	Wholesale	Manufact. Finished Inventories	Manufact. Input Inventories
1960: 1–1960: 4		−6.3	−1.7	−3.1	−6.3
1969: 3–1970: 4		−8.2	1.2	−0.4	−5.2
1973: 4–1975: 1		−16.0	−5.8	2.4	−13.2
1980: 1–1980: 2		3.6	1.9	−0.3	−4.1
1981: 3–1982: 4		−7.6	−2.3	−7.8	−11.1

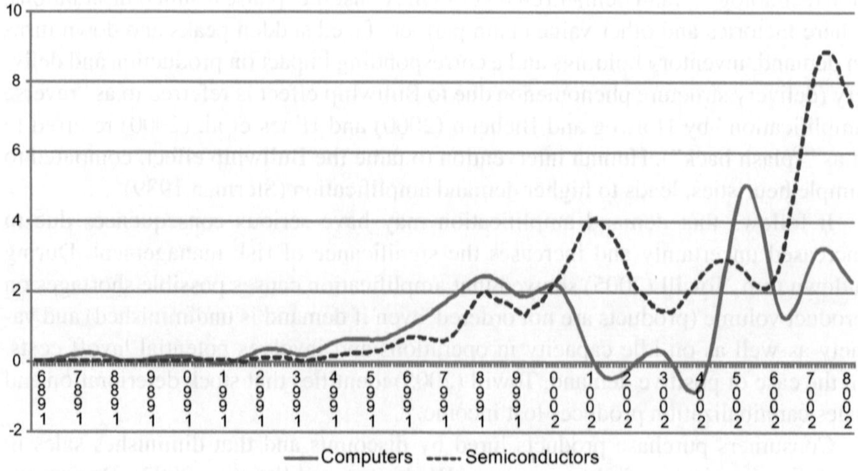

Fig. 10.1 Capacity addition change in US computer and semiconductor 1986–2008 (Federal Reserve 2008)

inventory positions seem to fluctuate, Albertson and Aylen (2003) argue that autoregressive forecasting models are able to forecast next period situation with a 50 % accuracy. While autoregressive techniques have been widely used in finance (and economics) in the past few decades, they may not have been applied or explored as decision support tools by supply chain planners or analysts in the area of supply chain management (Datta et al., 2007) or in other verticals (healthcare, energy).

Logic of demand amplification is evident in economic cycles (Forrester, 1976; Sterman, 1985). Order backlog, existing inventory holdings, amount of production, amount of employment and capacity additions were used in simulation trials to forecast different levels of economic cycles. In long-term changes, both Forrester (1976) and Sterman (1985) have emphasized the importance of capacity additions. A similar methodology has been used in maritime economics to estimate price level changes (Dikos et al., 2006) and investment cycle lengths in capital intensive industries (Berends and Romme, 2001). Hilmola (2007) has shown that capacity additions in US in semiconductors and computers industries may explain the behavior of stock market indices.

10.3 Beer Game and Role of Advanced Forecasting Methods

Forrester (1958) introduced a classical 4-stage beer-game simulation and revealed that demand information amplifies within the supply chain as we move further upstream. Figure 10.2 shows, customer demand is flat at 8 units per time period (it increased from 4 to 8 during time period of 100), but over-under reaction appears when supply chain is moved further with respect to time. Production orders spike to over 40 units per period, while waiting collapses to 0 units only 15 time units later. This occurs mostly due to time-delayed supply process in each stage, which is following make-to-stock (MTS) inventory principles (each phase has "target" for end-item inventory levels, which they try to reach with order algorithms).

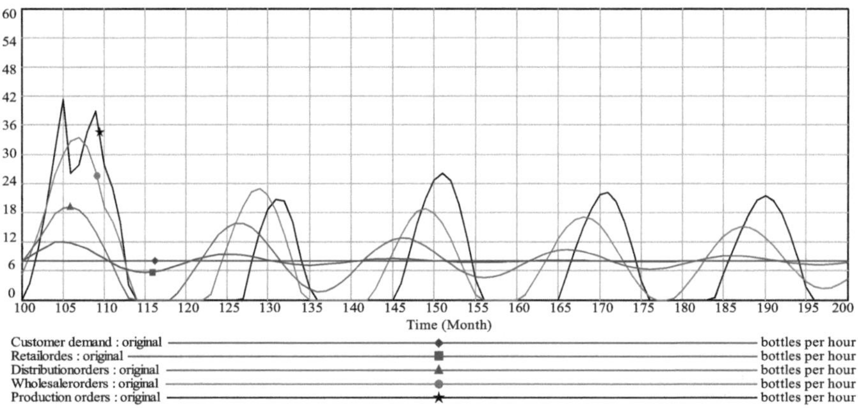

Fig. 10.2 Forrester effect (delay = 4, step-wise demand change from 4 to 8 units during time unit 100)

It may be observed from previous research that conventional forecasting methods do not reduce negative impact from demand amplification. As shown in Fig. 10.3, classical forecasting techniques such as exponential smoothed moving average (EWMA) only heightens demand amplification (highest value reaches above 50 units per time period) due to the assumption that all previous values should be used to predict demand. Although, use of EWMA may be justified under certain circumstances, the non-discriminatory or mandatory use of past data to predict future demand may often generate an undesirable over-reaction.

Figure 10.4 reveals implementation problems for sharing of demand information. Often different supply chain phases use different competing suppliers to gain cost efficiency and true demand is often confidential. Lam and Pestle (2006) describe 60 % of respondents in a survey (in China) indicating that their customers are not willing to exchange information. Sharing data about high demand periods could result in inflated purchase price if suppliers decide to form cartels. However, it has been shown in a signaling game theoretic approach that sharing of information increases total supply chain profit (Datta, 2004).

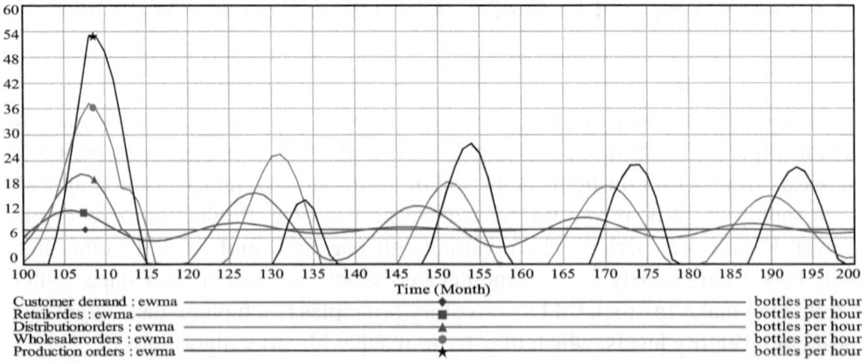

Fig. 10.3 Forrester effect in a supply chain as it tries to use EWMA at local level (0.5 weight) within original setting (delay = 4, step-wise demand change from 4 to 8 units during time unit 100)[1]

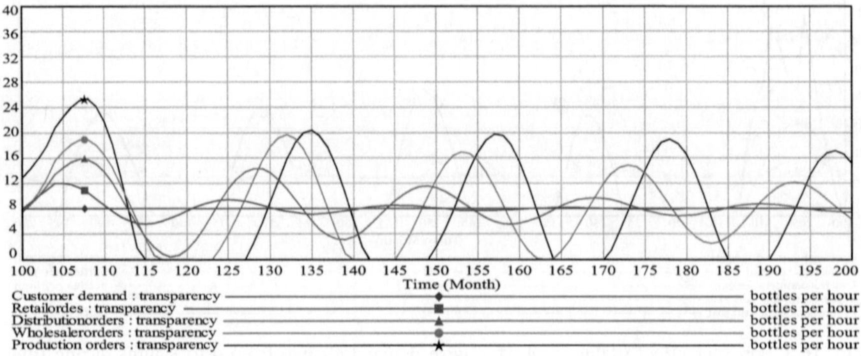

Fig. 10.4 Forrester effect in four stage supply chain, where we have transparency for the next stage (delay = 4, step-wise demand change from 4 to 8 units during time unit 100)

In previous and current work, we suggest that improving forecasting accuracy could be profitable by using advanced forecasting methods, such as autoregressive moving average (ARMA) models. Figure 10.5 shows that demand forecasting in an amplified environment may be completed by assigning a positive value for last observed demand and a negative co-efficient for older observations (for simplicity we have used lag of one and two).

Table 10.2 shows co-efficient of two lagging parameters are within the neighborhood of ARMA models built with a larger amount of data. However, the differences among different models are rather minimal.

Applying advanced forecasting models to tame the Bullwhip effect is challenging because it calls for process transformation (Zhao et al., 2002; Bayraktar et al., 2008). In Figure 10.6 the manufacturing unit reserves forecasted amount of inventory one

[1] Juha Saranen, Lappeenranta University of Technology, Finland

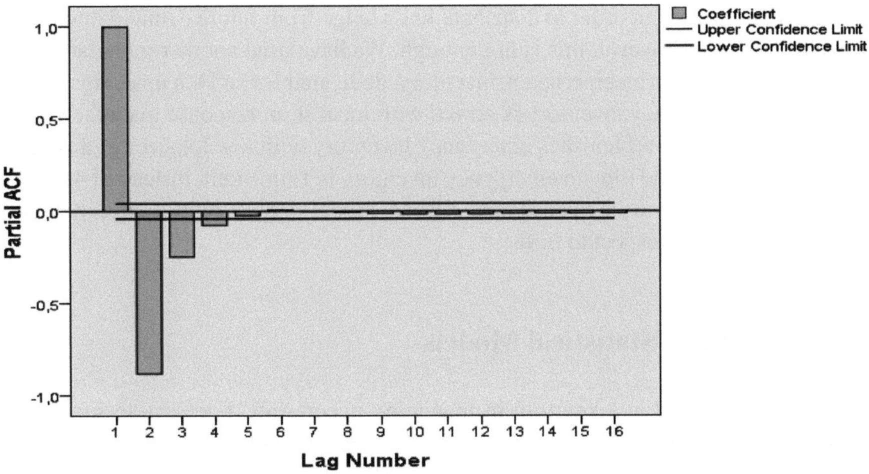

Fig. 10.5 Partial autocorrelation of four stage supply chain data from production phase (2401 observations from 300 time units)

Table 10.2 ARMA models built with 0.125 interval data from original beer-game setting for production phase (number of observations in integer time units given in parenthesis)

Number of observations	Co-efficient t −1	t −2	Goodness of fit R2 (%)	R2 (%) whole sample
100 (12.5)	1.936	−0.938	99.8	99.97
200 (25)	1.937	−0.939	99.9	99.97
300 (37.5)	1.937	−0.939	99.9	99.97
2401 (300.1)	1.918	−0.919	100.0	–

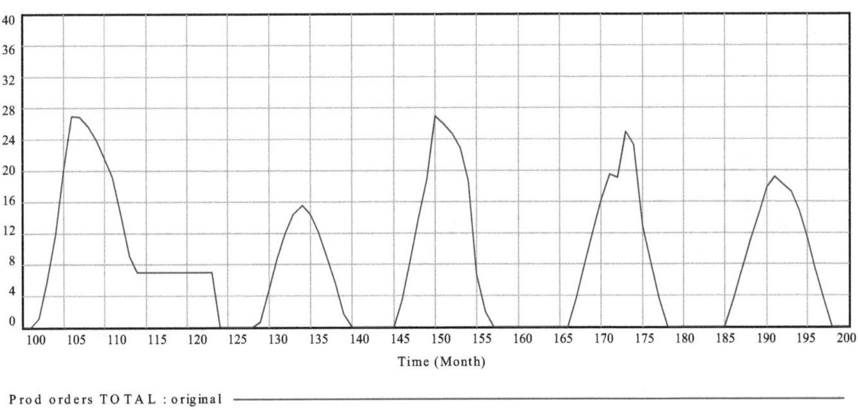

Fig. 10.6 Production orders as ARMA model of wholesaler demand is used in production orders with modifications on operating structure

period before-hand (in order to distribute knowledge from future demand into operative decisions). However, this is not enough. We have used another manufacturing unit, which serves as an emergency inventory, dedicated for sudden upswings of demand. This emergency inventory is served with local short response manufacturing, which continuously replenishes emergency inventory with low lot size (in this case lot size is 7 units, lead time to emergency inventory is 1 time unit, instead of 4). During simulation trials we explored how ARMA model may be built within dynamic environments with respect to time.

10.4 Advanced Statistical Models

Forecasting demand is a key tool in managing uncertainty. Forecast accuracy depends on the understanding and coverage of parameters as well as the accuracy of historic data available for each variable that may have an impact on the forecast or predictive analytics. The broad spectrum applicability of forecasting includes such diverse verticals as healthcare and energy [2] utilization (Datta 2008e, 2008f).

One of the assumptions in the Classical Linear Regression Models relates to homoskedasticity (homo $\approx$ equal, skedasticity $\approx$ variance or mean squared deviation (σ^2), a measure of volatility) or constant variance for different observations of the error term. Forecast errors are heteroskedastic (unequal or non-constant variance). For example, in multi-stage supply chains, the error associated with manufacturer's forecast of sales of finished goods may have a much larger variance than the error associated with retailer's projections (the assumption being that the proximity of the retailer to the end consumer makes the retailer offer a better or more informed forecast of future sales through improved understanding of end-consumer preferences). The upstream variability reflected in the Bullwhip Effect violates the basic premise of CLRM, the assumption of homoskedasticity. CLRM ignores the real-world heteroskedastic behavior of the error term ε_t and generates forecasts, which may provide a false sense of precision by underestimating the volatility of the error terms.

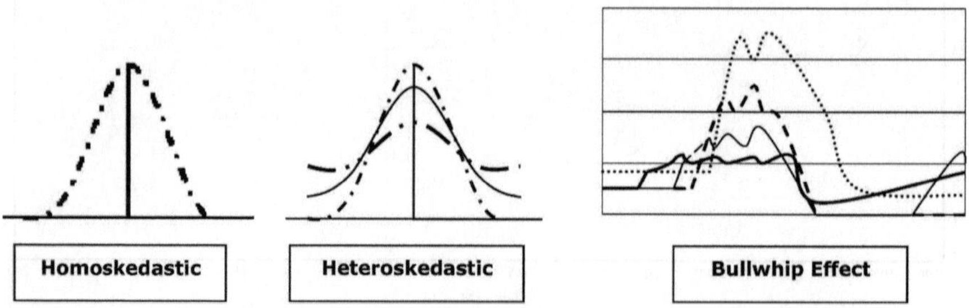

Fig. 10.7 Illustrations pf homoskedasticity, heteroskedasticity and the Bullwhip Effect

[2] See www.cids.ie/Research/DCSES.html and www.dcsenergysavings.com

In a homoskedastic distribution, observations of the error term can be thought of as being drawn from the same distribution with mean $= 0$ and variance $= \sigma^2$ for all time periods (t).

A distribution is described as heteroskedastic when observations of the error term originate from different distributions with differing widths (measure of variance). In supply chains, the variance of orders is usually larger than that of sales and the distortion increases as one moves upstream from retailer to manufacturer to supplier. Therefore, the assumption of heteroskedasticity, over time, seems appropriate as a characteristic that may be associated with demand amplification or the Bullwhip effect.

While variance of error term may change across cross sectional units at any point in time, it may also change over time. This notion of time varying volatility is frequently observed in financial markets and has been the driving force behind recent advancements in time series techniques. Instead of considering heteroskedasticity as a problem to be corrected (approach taken by CLRM practitioners in assuming homoskedasticity of error term), Robert Engle modelled non-constant time dependent variance (heteroskedasticity) using an autoregressive moving average (ARMA) technique.

ARMA consists of two components, an autoregressive (AR) compnent and a moving average (MA) component. AR is a technique by which a variable can be regressed on its own lagged values and thus link the present observation of a variable to its past history. For example, today's sales may depend on sales from yesterday and the day before. MA expresses observations of a variable in terms of current and lagged values of squared random error terms using a moving average.

Robert Engle used the MA approach to propose AutoRegressive Conditional Heteroskedasticity or ARCH to model the time varying volatility in a series. The 'conditional' nature of non-constant variance (heteroskedasticity) refers to forecasting of variance *conditional* upon the information set available. This MA representation was later generalized to an ARMA representation referred to as Generalized AutoRegressive Conditional Heteroskedasticity model or GARCH. The GARCH technique represents a parsimonious model than ARCH, while allowing for an infinite number of past error terms to influence current conditional variance.

These advances from econometrics may be developed into tools for forecasting and risks analytics with a broad spectrum of applications in business, energy, industry, security and healthcare (Datta, 2008d), as well as decision support systems and operations management, for example, supply chain management (Datta et al., 2007).

Impact of Real-Time High-Volume Data from Automatic Identification Technologies (AIT)

Tracking technologies evolved from the discovery of the RADAR at MIT in the 1940's. AIT is slowly beginning to impact information flow in the modern value chain network. Tracking products from manufacturers to retailers may have its origins in the 1970's with the introduction of the bar code to identify stock-keeping units (SKU). Now it is embracing AIT or auto id, for example, use of radio frequency identification (RFID). AIT makes it possible to electronically log product movement in the "digital" supply chain. This information may be available in real-time. However, the standards, such as the electronic product code (EPC), to capture unique identification of physical objects and processes (Datta, 2007a) calls for a paradigm shift (Datta, 2008c).

Since RFID updates reports every time an individual item or SKU moves from one stage to another stage in the supply chain, or when the item is sold, it is possible to determine the demand for an item in *real-time* rather than wait for batch updates or weekly or monthly buckets to generate a forecast. The granularity of the data from auto id systems may result in very high volume data which may reveal peaks and troughs of demand for the product, hourly or daily. This volatility is lost when data is aggregated in buckets or batch.

Extracting the value from this high volume near real-time data and deciphering the meaning of the implicit volatility may be a boon to business intelligence and predictive analytics, including forecasting.

Indeed many warehouses adopt an inventory policy of ordering products when stock levels fall below a certain minimum amount s and order up to a maximum amount S [the (s, S) policy]. With auto id it is possible to ascertain this at the instant the threshold is attained, thereby, eliminating the likelihood of out of stock (OOS). Hence, it follows that capitalizing on the increased volume of near real-time demand data from auto id may have profound impact on supply chain forecasting.

However, current software with its CLRM engines and clustering approach, does not improve forecasts even with high volume data. The assumption of error terms implicit in CLRM limits the gains in forecast accuracy from high volume data and fails to show return on investment (ROI) from adoption of (new) auto id tools.

It is our objective to justify why deployment of new tools, for example, auto id, calls for adoption of new techniques for data analytics, for example, advanced techniques from financial econometrics. It is safe to state that new streams of data emerging from a multitude of sources, for example, auto id and sensors, cannot yield value or ROI if used in conjunction with archaic software systems running ancient forms of analytical engines that are typically CLRM based, at least, in the forecasting domain.

To extract decisionable information from high volume near real-time auto id and sensor data, the use of techniques like GARCH deserves intense exploration. The fact that GARCH may be a clue to generating ROI from auto id and RFID data is no accident because GARCH requires high volume data to be effectively utilized and generate results with higher accuracy levels. Hence, this convergence of auto id data with tools from econometrics may be an innovative confluence that may be useful in any vertical in any operation including security and healthcare, as well as obvious and immediate use in supply chain management. Datta et al. (2007) and this chapter, has attempted to highlight how to extract the advances in econometrics from the world of finance and generalize their valuable use in decision systems, with a broad spectrum of general applications.

Evolution of such a tool may help analysts and planning managers since a key concern of any manager is the accuracy of the predictions on which their budget is based. The proper allocation of resources for acquisition of personnel and equipment has long been plagued by errors in traditional forecasting. A part of the answer to this problem may be latent in the potential for the combined use of VAR (vector autoregression) with other forecasting techniques. The primary VAR model best suited for the planning function in resource allocation is the standard GARCH model. It is well suited for pragmatic studies involving supply chain, army personnel requirements and defense equipment requirements. Any system that may be modeled using time-series data, may explore how to include GARCH based on the error correction innovation that may improve forecasts. Forecast accuracy of GARCH model may be quantified in a number of different ways. Traditional methods are:

1. Mean Square Error.
2. Mean Average Percentage Error.
3. Aikiki Information.

Once the forecast is developed, the accuracy can be measured by comparing the actual observed values with the predicted values. If the collected data falls within the confidence interval of the forecast model, then the model provides a good fit for the system. Although GARCH is useful in forecasting it is important to realize that it was designed to model volatility and can be applied to positive series but not to economic series.

GARCH Proof of Concept Demonstrated Using an Example from Spare Parts Inventory Management

Although GARCH models have been almost exclusively used for financial forecasting in the past, we propose (Datta et al. 2007b) that with appropriate modifications, it may be applicable to other areas. An operation exhibiting volatility may benefit from VAR-GARCH in addition to other techniques, either in isolation or in combination. It is known that in times of conflict military supply chains experience spikes in demand. These spikes and troughs occur over a short period of time and result in losses due to OOS (out of stock) or surplus. GARCH may minimize forecast errors and fiscal losses in some of the following domains:

1. Cost of personnel, supplies, support;
2. Planning, programming and budgeting;
3. Defense program and fiscal guidance development;
4. Force planning and financial program development.

In one preliminary study (supported by the US DoD, Institute for Defense Analysis, Washington DC and also mentioned in Datta et al., 2007) the spare parts supply chain of a military base was examined for inefficiencies. The data for a 9 month period was collected for a spare part for a military vehicle (HumV). Because the

US Department of Defense affixes RFID tags on some spare parts from some of its suppliers, the hourly auto id demand data was available for analysis. The historical data was used to develop CLRM and GARCH (1,1) model. In this case, the linear regression model was found to have Mean Square Error (MSE) of 0.20 (20 %). By comparison, the GARCH (1,1) model produced a MSE of 0.06 (6.7 %). These results are encouraging and the US DoD case lends credibility for exploration of GARCH in forecasting analytics. Although this finding is promising, it needs to be repeated with other forms of data and subjected to rigorous mathematical analysis.

Financial Profit from Application of GARCH Technique in Retail Inventory Management

A pilot implementation using GARCH in a commercial supply chain has been undertaken with real-world retail data from a major US retailer. Preliminary results reveal that using GARCH as a forecasting technique offers some advantages (even with limited data volume) compared to CLRM and ARMA. The retail data are from office supply products for business and home customers. Thousands of product lines are sourced from different suppliers, globally. In this study, nine different SKU's from three different product classes were chosen for analysis. It is not known whether some of the products may suffer from seasonality effects. For each of the 9 products the historical demand data is available for 70 continuous weeks. 52 weeks of history was used to develop projections of demand variability for the subsequent 18 weeks. This projection was compared to the actual observations of demand variability over the 18 week test horizon. Three different techniques were used to forecast the standard deviation. The methods used were CLRM, ARMA and GARCH.

Figure 10.8 shows the performance of each of the three forecasting techniques based on retail data on the 9 products. The error of the forecasted standard deviation is calculated as an average over the final 16 weeks of forecast data. CLRM is outperformed by ARMA and GARCH models for almost all SKU's.

The results suggest that GARCH may be better (or as good as) across different SKU's. GARCH outperforms ARMA for a number of products. For the eight favorable tests, an average improvement of 800 basis points was observed. That translates to 1.1 % in-stock improvement from a preliminary application of GARCH. In this real-world case, that amounts to about $ 13 million in additional revenue in terms of recovered lost sales. The product-dependent variability of GARCH performance may be linked to seasonality or other factors (accuracy of input data). Further testing with granular data (hourly or daily) and higher volume data per SKU may increase the accuracy and benefits from using the GARCH technique in forecasting. It may be apparent that systemic use of GARCH type techniques in forecasting, therefore, may substantially improve corporate profit.

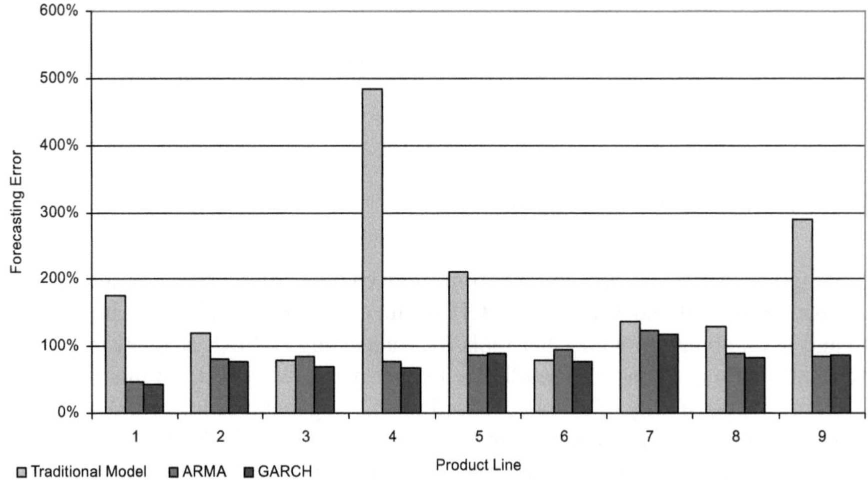

Fig. 10.8 Classic Linear Regression Model (CLRM) is almost always outperformed by ARMA and GARCH

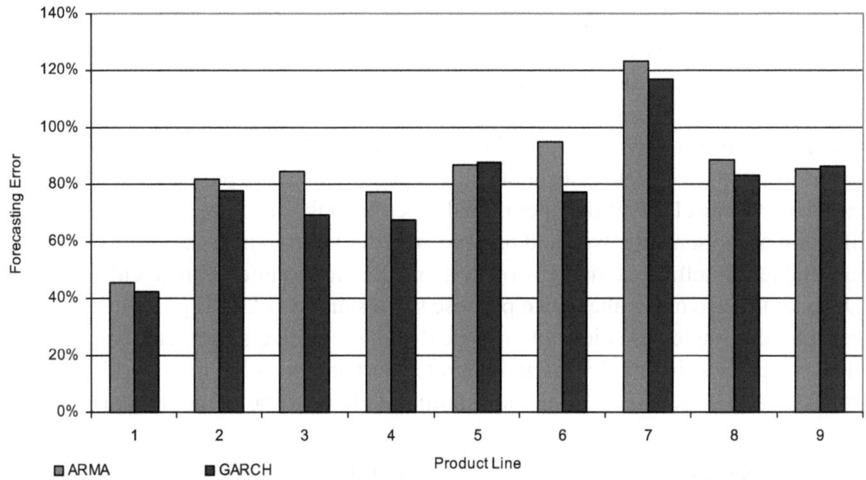

Fig. 10.9 Comparing autoregressive techniques: ARMA vs. GARCH (same data set but excluding CLRM)

It is not difficult to extrapolate that high volume item level retail sales or inventory data (per minute or by the hour) may be available with the diffusion of item level tagging using RFID or radio frequency identification tags. The volume of the data may increase exponentially if embedded sensors are deployed to enhance security and/or detect movement of any physical object from any location. Businesses dealing with short life cycle products (electronics, semi-conductors industries) may explore how these advanced techniques may help to reduce the volatility of supply-demand since the ability to re-address sales or marketing issues are often limited if the shelf-life of the product is merely a few months (laptops, MP3 players, cell phones).

10.5 Temporary Conclusion

Making sense of data may benefit from high volume data acquisition and analysis using GARCH and VAR-MGARCH (Datta et al., 2007) techniques in addition to and in combination with other tools for forecasting and risk analysis in diverse verticals that may span from healthcare to energy (Datta, 2008e). In this work, we explored the possibility of using advanced forecasting methods in context of supply chains and demonstrated financial profitability from use of the GARCH technique. It remains unexplored if concomitant business process transformation may be necessary to obtain even better results. The proposed advanced forecasting models, by their very construction require high volume data. Availability of high volume data may not be the limiting factor in view of the renewed interest in automatic identification technologies (AIT) that may facilitate acquisition of real-time data from products or objects with RFID tags or embedded sensors. It is no longer a speculation but based on proof that use of advanced forecasting methods may enhance profitability and ICT investments required to acquire real-time data may generate significant return on investment (ROI). However, understanding the "meaning" of the information from data is an area still steeped in quagmire but may soon begin to experience some clarity if the operational processes take advantage of the increasing diffusion of the semantic web and organic growth of ontological frameworks to support ambient intelligence in decision systems coupled to intelligent agent networks (Datta, 2006). To move ahead, we propose to bolster the GARCH proof of concepts through pilot implementations of analytical engines in diverse verticals and explore advanced forecasting models as an integrated part and parcel of real-world business processes and systems including the emerging field of carbonomics (Datta, 2008f).

Acknowledgements The corresponding author (SD) is deeply indebted to Prof. Clive Granger (UCSD) for his guidance, critical reading and review of the manuscript. The author (SD) is grateful to Professor Robert Engle (NYU) for his advice. The detailed editing of this document was accomplished by Brigid Crowley at the School of Science and Computing at ITT, Ireland. Comments from General Paul Kern of The Cohen Group (Washington DC) and Dr. Stan Horowitz of the Institute for Defense Analysis (Washington DC) were very helpful to the corresponding author (SD). The encouragement from Mr Kazunori Miyabayashi, Senior Vice President of Hitachi (Tokyo, Japan) and Professor Joseph Sussman, MIT is acknowledged. This work was partially supported by the members and sponsors of the MIT Forum for Supply Chain Innovation, Massachusetts Institute of Technology (Cambridge, Massachusetts) and the Centre for Innovation in Distributed Systems.

References

Arvis J.-F., Mustra M.A., Panzer J., Ojala L., and Naula T. (2007). *Connecting to Compete – Trade Logistics in the Global Economy*. World Bank Publications, New York

Albertson K. and Aylen J. (2003). Forecasting the behavior of manufacturing inventory. *Int J Forecasting* 19:299–311

Bayraktar E., Lenny Koh S.C., Gunasekaran A., Sari K., and Tatoglu E. (2008). The role of forecasting on Bullwhip effect for E-SCM applications. *International Journal of Production Economics* 113:193–204

Berends P.A.J. and Romme A.G.L. (2001). Cyclicality of capital intensive industries: A system dynamics simulation study of the paper industry. *Omega: International Journal of Management Science* 29:543–552

Boyle E., Humphreys P., and McIvor R. (2008). Reducing supply chain environmental uncertainty through e-intermediation: An organization theory perspective. *International Journal of Production Economics* 114:347–362

Chatfield C. and Yar M. (1988). Holt-Winters forecasting: Some practical issues. *The Statistician* 37(2):129–140

Hong C., Frank M.Z., and Wu Q.Q. (2005). What actually happened to the inventories of American companies between 1981 and 2000? *Management Science* 51(7):1015–1031

Coase R.H. 1937. "The Nature of the Firm" Economica N.S. 4:386–405. Reprinted in Williamson O.E. and Winter S., eds., 1991. The Nature of the Firm: Origins, Evolution, Development. NY. Oxford University Press, pp. 18–33

Coase R.H. (1960). The problem of social cost. *J Law Econ* 3:1–44

Coase R.H. 1972. Industrial Organization: A Proposal for Research in Fuchs V.R., ed., Policy Issues & Research Opportunities in Industrial Organization. NY: National Bureau of Economic Research, pp. 59-73

Coase R.H. 1992. The Institutional Structure of Production. *American Economic Rev* 82:713–719

Cooper D.P. and Michael T. (2005). Supply chain integration via information technology: strategic implications and future trends. *International Journal of Integrated Supply Management*, 1(3):237–257

Dapiran P. (1992). Benetton: global logistics in action. *International Journal of Physical Distribution & Logistics Management* 22(6):7–11

Datta S., *et al.* (2004). "Adaptive value networks: emerging tools and technology as catalysts″ in Y.S. Chang, H.C. Makatsoris, and H.D. Richards (eds). *Evolution of Supply Chain Management: Symbiosis of Adaptive Value Networks and ICT*. Kluwer Academic Publishers, Amsterdam-Boston

Datta S. (2006). *Potential for improving decision support catalysed by semantic interoperability between systems*. Working Paper, Engineering Systems Division, Massachusetts Institute of Technology. http://esd.mit.edu/WPS/esd-wp-2006-10.pdf & http://dspace.mit.edu/handle/1721.1/41903

Datta S., Granger C.W.J., Barari M., and Gibbs T. (2007). Management of supply chain: an alternative modelling technique for forecasting. *Journal of the Operational Research Society* 58:1459–1469 http://dspace.mit.edu/handle/1721.1/41906 and http://esd.mit.edu/WPS/esd-wp-2006-11.pdf

Datta S. (2007a) Unified Theory of Relativistic Identification of Information in a Systems Age: *Convergence of Unique Identification with Syntax and Semantics through Internet Protocol version 6 (IPv6)*. Engineering Systems Division Working Paper Series, Massachusetts Institute of Technology http://esd.mit.edu/WPS/2007/esd-wp-2007-17.pdf & http://dspace.mit.edu/handle/1721.1/41902

Datta S. (2008a). *Forecasting and Risk Simulation: Proposed Analytical Tool*. Working Paper, Engineering Systems Division, Massachusetts Institute of Technology http://dspace.mit.edu/handle/1721.1/41913

Datta S. (2008b) WiFi Meet FuFi: Disruptive Innovation in Logistics Catalysed by Energy. *International Journal of Electronic Business Management* 6 117-119 http://dspace.mit.edu/handle/1721.1/41897

Datta S. (2008c) Auto ID Paradigm Shifts from Internet of Things to Unique Identification of Individual Decisions in System of Systems. MIT ESD Working Paper Series and *Supply Chain Europe*, 2008 http://esd.mit.edu/WPS/2008/esd-wp-2008-09.pdf and http://dspace.mit.edu/handle/1721.1/41900

Datta S. (2008d) Will Nano-Butlers Work for Micro-payments? Innovation in Business Services Model may Reduce Cost of Delivering Global Healthcare Services. MIT Engineering Systems Division WP http://esd.mit.edu/WPS/2008/esd-wp-2008-17.pdf and http://dspace.mit.edu/handle/1721.1/41901

Datta S. (2008e) Tutorial on Sensor Networks in Energy Management Solution http://dspace.mit.edu/handle/1721.1/43941

Datta S. (2008f) Nano-sensoromics: Is it Conceptually Similar to Carbonomics? http://dspace.mit.edu/handle/1721.1/439415

Deutsche Post (2007). Annual Report of 2007. Germany

George D., Marcus H.S., Papadatos M.P., and Papakonstantinou V. (2006). Niver Lines: A system-dynamics approach to tanker freight modeling. *Interfaces*, 36(4):326–341

Fisher M.L., Hammond J.H., Obermeyer W.R., and Raman A. (1994). Making Supply Meet Demand In an Uncertain World. *Harvard Business Review* 72(3):83–93

Fraiman N. and Singh M. (2002). Zara: Case Study. Columbia University School of Business, New York

Forrester J.W. (1958). Industrial dynamics – a major breakthrough for decision makers. *Harvard Business Review* 36(4):37–66

Forrester J.W. (1976). Business structure, economic cycles, and national policy. *Futures* 8(3):195–214

Hilletofth P. and Hilmola O.-P. (2008). Supply chain management in fashion and textile industry. *International Journal of Services Sciences* 1(2):127–147

Hilmola O.-P. (2007). Stock market performance and manufacturing capability of the fifth long-cycle industries. *Futures* 39(4):393–407

Hilmola O.-P. and Szekely B (2008). *Railways, Airports and Sea Container Operators as Publicly Listed Companies – Financial Performance and Shareholder Value Creation Perspective*. Lappeenranta University of Technology, Department of Industrial Management. Research Report 196

Hines P., Holweg M., and Sullivan J. (2000). Waves, beaches, breakwaters and rip currents – A three-dimensional view of supply chain dynamics. *International Journal of Physical Distribution & Logistics Management* 30(10):827–846

Holweg M. and Bicheno J. (2000). The reverse amplification effect in supply chains. In: Katayama, H. (ed.): *Global Logistics for the New Millennium – 5th International Symposium on Logistics*, pp. 549–554. Tokyo

Kros J.F., Falasca M., and Nadler S.S. (2006). Impact of just-in-time inventory systems on OEM suppliers. *Industrial Management and Data Systems* 106(2):224–241

Lam J.K.C. and Postle R. (2006). Textile and apparel supply chain management in Hong Kong. *International Journal of Clothing Science and Technology* 18(4):265–277

Lee H.L., Padmanabhan P., and Whang, S (1997). The Bullwhip Effect in Supply Chains. *MIT Sloan Management Review* 38:93–102

Lowson R. (2001). Analysing the effectiveness of European retail sourcing strategies. *European Management Journal* 19(5):543–551

Moon M.A., Mentzer J.T., and Thomas D.E. (2000) Customer demand planning at Lucent Technologies – A case study in continuous improvement through sales forecast auditing. *Industrial Marketing Management* 29:19–26

Oliver R.K. and Webber M.D. (1982) Supply-chain management: logistics catches up with strategy. In Christopher M (ed.): *Logistics: Strategic Issues*, Chapman & Hall, London

Ramey V.A. (1989). Inventories as factors of production and economic fluctuations. *The American Economic Review* 79(3):338–354

Reyes P.M. and Frazier G.V. (2007) Radio frequency identification: Past, present and future business applications. *International Journal of Integrated Supply Management* 3(2):125–134

Reyes P.M., Frazier G.V., Prater E.L., and Cannon A.R. (2007) RFID: The state of the union between promise and practice. *International Journal of Integrated Supply Management* 3(2):192–206

Stratton R. and Warburton R.D.H. (2003) The strategic integration of agile and lean supply. *International Journal of Production Economics* 85(2):183–298

Sterman J.D. (1985) An integrated theory of the economic long wave, *Futures* 17(2):104–131

Sterman J.D. (1989). Modeling managerial behavior: Misperceptions of feedback in a dynamic decision making experiment. *Management Science* 35(3):321–339

Towill D.R. (2005) The impact of business policy on bullwhip induced risk in supply chain management. *International Journal of Physical Distribution and Logistics Management* 35(8):555–575

United Nations (2007). *Regional Shipping and Port Development Strategies – Container Traffic Forecast 2007 Update*. Economic and Social Commission for Asia and the Pacific

United Nations (2005) *Regional Shipping and Port Development Strategies (Container Traffic Forecast)*. Economic and Social Commission for Asia and the Pacific

Warburton R.D.H. and Stratton R. (2002). Questioning the relentless shift to offshore manufacturing. *Supply Chain Management: An International Journal* 7(2):101–108

Wright D. and Yuan X. (2008) Mitigating the bullwhip effect by ordering policies and forecasting methods. *International Journal of Production Economics* 113:587–597

Zhao X., Xie J., and Lau R.S.M. (2001) Improving the supply chain performance: Use of forecasting models versus early order commitments. *International Journal of Production Research* 39(17):3923–3939

Zhao X., Xie J., and Leung J. (2002) The impact of forecasting model selection on the value of information sharing in a supply chain. *International Journal of Production Economics* 142:321–344

Chapter 11
Supply Chain Risk Management: Annotation of Knowledge Using a Semi-Structured Knowledge Model

Chun-Che Huang and Tzu-Liang (Bill) Tseng

Abstract In order to increase the effectiveness and performance of supply chain networks, firms are working towards improving Supply Chain Risk Management (SCRM) knowledge. Appropriately managing SCRM knowledge can result in lower costs, fewer operational disruptions and better customer satisfaction due to supplier uncertainty has been eliminated. The lack of flexible- and well-structured knowledge representation has made the integration of SCRM knowledge activities difficult because these knowledge activities are heterogeneous in nature. Semantic heterogeneity is the main problem of SCRM knowledge representation. Annotation is one of the approaches to solve this problem. Annotation adds semantic metadata on web documents and proposes machine-understandable metadata for integration of heterogeneous format documents. This book chapter proposes a semi-structured knowledge (SSK) model to represent SCRM knowledge and uses Resource Description Framework (RDF) and Resource Description Framework System (RDFS) as metadata languages to annotate semantic metadata. This proposed annotation process integrates heterogeneous and un-structured SCRM documents. This solution approach shows great promise for knowledge representation/sharing in the heterogeneous SCM environment.

Key words: Semi-structured knowledge model; Semantic heterogeneity; Ontology; Annotation process; Supply chain risk management

11.1 Introduction

Managing supply chains in today's competitive world is increasingly challenging. Greater the uncertainties in supply and demand, globalization of the market, shorter and shorter product and technology life cycles, and the increased use of manufacturing, distribution and logistics partners have resulted in complex international supply network relationships, have led to higher exposure to risks in the supply chain

(Christopher et al., 2002). The complexity and uncertainty within a supply chain can also increase the "chaos" risks within the supply chain. These chaos effects result from over reactions, unnecessary interventions, mistrust, and distorted information throughout a supply chain (Childerhouse et al., 2003). Consequently, global competition and technological change have motivated improved risk management in supply chains. Risk management in supply chains consists of risk sharing, control and prevention, and financial instruments to negate the effects of the supply chain risks and their capital consequences (Lee, 2004; Hallikas et al., 2004; Tang, 2006). With the increasing emphasis on supply chain vulnerability, effective tools for analyzing and understanding appropriate supply chain risk management are now attracting much attention.

To survive in a very competitive market today, firms are required to efficiently operate the core knowledge domain (Badaracco, 1990; Dieng et al., 1999) and focus on high-value and applicable knowledge. The lack of flexible and well-structured knowledge representation has made the integration of Supply Chain Risk Management (SCRM) knowledge activities difficult because these knowledge activities are heterogeneous in nature. (Nonaka and Takeuchi, 1995). Therefore, how to make representation of SCRM knowledge consistent and flexible is critical (Cui et al., 2001). A representation model is required to present knowledge with a flexible and well-structured format in order to store and reuse them (Malone, 2002; Woodworth and Kojima, 2002; Nabuco et al., 2001). Furthermore, access to relevant and accurate information/knowledge is becoming increasingly complex due to the databases and knowledge bases that are distributed, diverse, and dynamic. Resolving the heterogeneity of SCRM knowledge documents from the various hetero-purpose knowledge-based systems has become a crucial problem. How to efficiently access and retrieve these hetero-format knowledge documents, specifically solving the semantic heterogeneity, is an imperative issue (Decker et al., 1999). A process by annotation to transform these heterogeneous SCRM knowledge documents into a flexible and well-structured SCRM knowledge is required.

SCRM related documents exist in various formats according to different sources, such as documents in text, paper, and audio formats. Some documents are archived as images. Based on the literature review, two issues are focused in SCRM knowledge representation. One issue is improving the document representations while the other issue is refining user queries. Note that this paper focuses on the first issue. SCRM documents include the combination of qualitative and quantitative parts, for example, tier I and tier II suppliers documents, documentation related to natural hazards, terrorism, pandemics, data security, demand variability and supply fluctuations, etc. Current research on SCRM document representation is not sufficient to improve efficiencies in the knowledge representation. For example, documents are indexed by relatively few terms compared to using the whole index space and the SCRM document classification rules are not considered in this indexing task. A classification rule that is related to the SCRM document representation approach is required to effectively index the documents. Annotation is a process that makes knowledge sources, including Web pages, un-structured text documents etc., on the basis of a formal description of their contents (Decker, 2002; Li et al., 2001).

That is to make the knowledge sources understandable for computer devices. These sources should be annotated with meta-data markups in an annotation process. In other words, annotating the contents of knowledge sources with meta-data markups is one promising method to make them understandable for machines in the semantic web environment. In the annotation domain, metadata (i. e., data about data) is one kind of semantically information allows the web to describe semantic properties about some given knowledge contents of knowledge documents (Kashyap and Shet, 1996; Decker, 2002; Fensel et al., 2000). According to definition of annotation, annotation can help with improving the document representations.

Annotation research attempts to add semantic metadata in web documents and propose machine-understandable metadata (Kashyap and Shet, 1996) for integration of a heterogeneous format document from web knowledge sources in a structured model. However, little research focuses on annotating metadata from a knowledge content perspective, such as annotation for knowledgeable workers rather than computer, and modeling SCRM knowledge documents through different dimensions in order to make the knowledge learner recognize the usefulness of designated knowledge. To operate the annotated metadata, currently three major knowledge-based systems extensively used in semantic web: Simple HTML Ontology Extensions (SHOE) (Heflin and Hendler, 2000), Ontobroker (Decker et al., 1999), and WebKB (Martin and Eklund, 1999). All of these studies rely on knowledge in HTML language. They all started with providing manual mark-up by editors. However, experiences of Erdmann et al. (2000) have shown that mark-up knowledge in these three systems yields extremely poor results that contain syntactic mistakes, improper references, and all of the problems illustrated in the scenario section.To avoid redundancy of the use of meta-datum, an approach to coordinate and manage meta-datum in the metadata repository and benchmark ontology is definitely required.

This book chapter is based on a semi-structured knowledge (SSK) model (Huang and Kuo, 2003) to presents SCRM knowledge. It also presents tools and techniques for decision making related to supply chain risk and vulnerability and uses Resource Description Framework (RDF) and Resource Description Framework System (RDFS) as metadata languages to annotate SCRM knowledge documents which are in unstructured heterogeneous format semantically. Moreover, it proposes an annotation process to synthesize heterogeneous un-structured documents form heterogeneous SCRM knowledge sources and presents interaction between SSK and benchmark ontology. Note that the benchmark ontology in this chapter refers to the specific ontology in use. The remainder of this book chapter is organized as follows: Sect. 11.2 describes the generation and contents of semi-structured knowledge and the SCRM semi-structured knowledge represented with RDF and RDFS. Sect. 11.3 proposes the annotation process to transform the original heterogeneous documents into RDF and RDFS documents using the SSK model. Sect. 11.4 presents interaction of annotated SCRM knowledge documents with the benchmark ontology and Sect. 11.5 concludes this book chapter. The main contribution of this book chapter is that the proposed approach develops a standard way to exchange knowledge and it can be applied in any general SC domain through development of a SSK model and annotation framework. Moreover, in order to manage SC risk effectively, web based SCRM and its documentation transitions and representations play an important role. Particularly the heterogeneity problem among documentations is required to be solved. The heterogeneity issue is resolved by the annotation approach in perspective of SCRM knowledge content is understandable by both machines and knowledgeable workers. The solution approach demonstrates great promise for effective knowledge sharing in SCRM.

11.2 The Generation and Representation of Semi-Structured Knowledge

11.2.1 The Generation of Semi-Structured Knowledge

Semi-structured knowledge (SSK) is defined as collection of knowledge resulting from Knowledge Management (KM) activities in organizations and constructed by the six dimensions of Zachman framework (5W1H: who, what, when, why, where and how) (Huang and Kuo, 2003). The Zachman framework represents the perspectives and dimensions of knowledge contents in the form of a matrix, with the perspective representing the rows, e.g. goals, and the dimensions representing the columns, including: Entities (What? Things of interest); Activities (How? In what manner or way; by what means); Locations (Where? Places of interest); Individual (Who? Individuals and organizations of interest); Times (When? Things occur) and Motivations (Why? Reasons and rules).

Notations
D_A: annotation description,
D_B: basic description,
D_R: relationship description,
I_G: general information,
I_S: solution approach information,
K_A: annotated semi-structured knowledge,
K_O: organizational semi-structured knowledge,
K_P: knowledgeable workers or problem solver,
K_T: understandable knowledge.

SSK contains large amounts of solution approach information, domain knowledge, and know-how and the environment impact on organizations closely. It is generated through a series of transformation processes (Fig. 11.1), which are triggered by problematic events (e.g., inventory is too high). Every step of this process involves technical and managerial issues. At the beginning, when a problematic event occurs, the event initiates an operation to identify requirements of users and starts a problem-solving process.

(1) If the problem can be solved in an organization internally, related knowledge elements (concepts, properties, and instances) are retrieved from the bench-

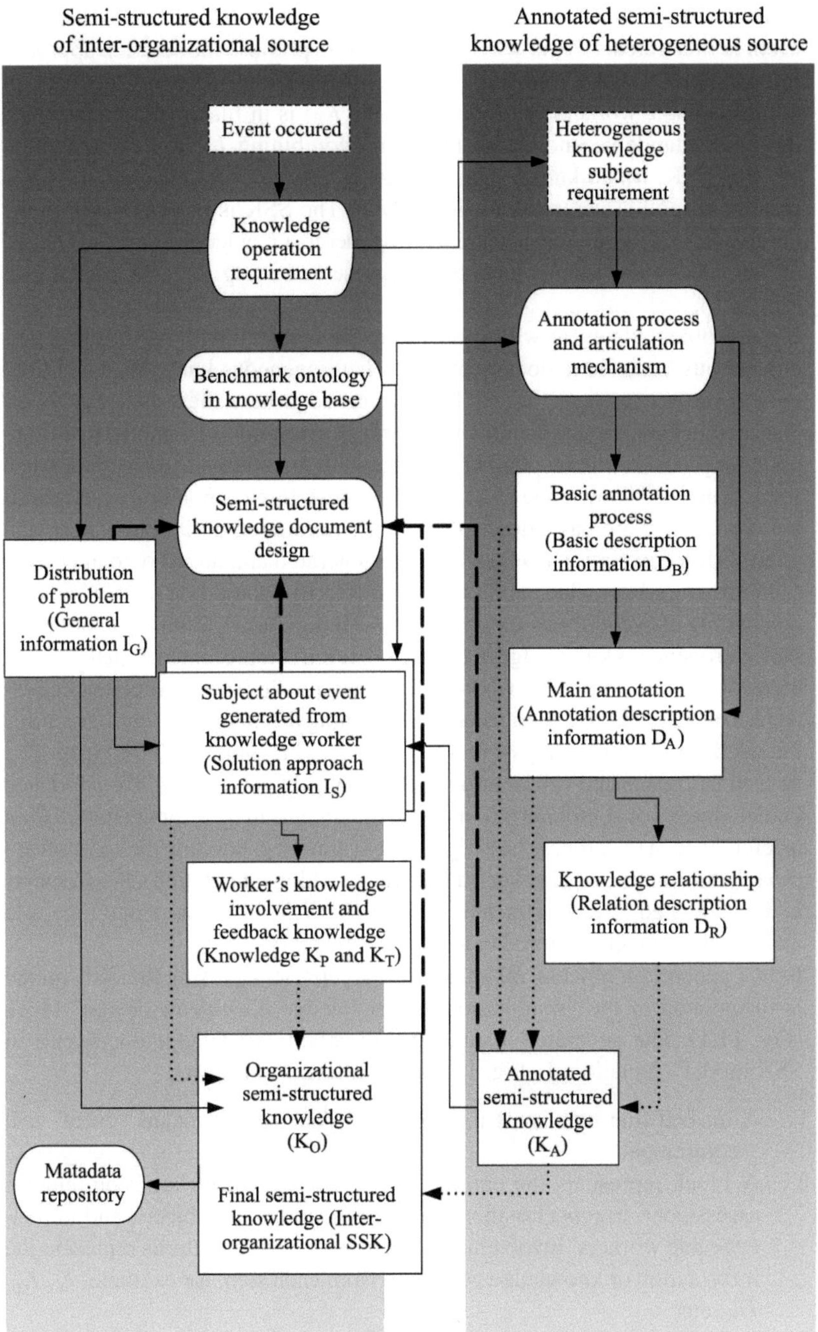

Fig. 11.1 Flow diagram of SSK generation

mark ontology as follows: (i) general information (I_G) of the problematic event is illustrated, (ii) knowledgeable workers propose the solution approach information (I_S) after inferring to the benchmark ontology, (iii) feedback of knowledgeable workers or problem solver (K_P) is in fusion and transformed into understandable knowledge (K_T), (iv) combining I_G, K_P, I_S, *and* other existing SSK, *which* knowledgeable worker may refer to, generate the organizational semi-structured knowledge (K_O).The SSK is produced within the organization without considering heterogeneous knowledge sources. That is, all knowledge documents have been represented using the SSK model (see Fig. 11.1).

(2) If the problematic event with its I_G needs be resolved with aid of other heterogeneous knowledge documents, then heterogeneous knowledge solution-related subject (also in Fig. 11.1) is identified, where these heterogeneous/unstructured knowledge documents are transformed into SSK model by the annotation process. The resulted knowledge with annotation is called annotated semi-structured knowledge (K_A). In this process, three annotation descriptions are used. There are basic description (D_B), annotation description (D_A), and relationship description (D_R), which are generated and added onto annotated semi-structured knowledge (K_A). K_A is primarily focused on the subject corresponding to event's reason (why) and resolution (how). With the annotation, the heterogeneous knowledge in the organization is represented with the six dimensions (what, who, why, when, where, and how) of solution-related knowledge. The final SSK is generated after combining the annotated semi-structured knowledge (K_A), the analysis reasons (why) and resolutions (how) from K_A, as well as the general information I_G from K_O. In some cases, the annotated knowledge is not combined into the final documents in SSK model directly but is referenced in the process (see Fig. 11.1). It aims at generating the solution approach information (I_S). In such a case (e. g., adding annotation), the elements in the annotated SSK is extracted and mapped to the benchmark ontology, and annotated document is stored in metadata repository.

In this generation process, all knowledge is represented using the SSK model is formulated in the "semi-structured knowledge documents design" block (Fig. 11.1). The generated documents are called "knowledge documents in SSK model". Notations in Fig. 11.1 are illustrated as follows:

1. A dashed line represents the source of data, which means "event" and "requirement".
2. A block represents the entities in the process. There are requirements assessment, events classification, identifications of subjects, and knowledgeable workers' involvement. The contents in parenthesis represent the information or knowledge generated from each step, for example, I_s, K_p, D_R, etc.
3. An oval block represents data storage and processes.
4. Three different types of lines are used in the diagram. The solid line illustrates the direction of the process, while the long dash line points out

the requirements of semi-structured knowledge. The dot dash line shows the steps to construct semi-structured knowledge. As shown above, semi-structured knowledge includes general information, solution approach information, and feedback from knowledgeable workers (called feedback knowledge) and annotated semi-structured knowledge (K_A). Fig. 11.1 represents the conceptual framework of SSK generation. The detailed models related to this framework can be referred in Huang and Kuo (2003).

11.2.2 Semi-Structured Knowledge Representation

A better presentation of SSK is proposed by using RDF and RDFS in this paper. Since the RDF language is developed based on XML and relies on the support of XML (URL RDF), the features of the XML are also inherited by RDF, such as principles of structure, extensibility, self-description, and separation between data and model. RDF contains, not only outstanding efficiency in data management and exchange, but also effectively presents every kind of knowledge in network, and provides a new frame for knowledge management (URL RDF). Even more important, the global goal of RDF is to define a mechanism for describing resources that makes no assumptions about a particular application domain, but defines (a priority) the semantics of any application domain (Fensel et al., 2000). The characteristic of RDF allows semi-structured knowledge to appear more domain-natural and acceptable in the future Web-based inter-organization.

Figure 11.2 shows the contents of the SSK in XML format, rather than RDF and RDFS format. The left side in Fig. 11.2 presents SCRM SSK, which includes general information, subject/solution approach information, and feedback knowledge, where the association rule is applies as a solution approach. The right side in the figure shows the semi-structured knowledge with annotated knowledge. In Fig. 11.2, the notation K_A is semi-structured annotation knowledge which is composed of the following information terms: D_B: basic description information; D_A: annotation description information of solution-related subject; and D_R: relation description information. WhereK_A represents the formal semi-structured knowledge after the annotation process and K_O represents the manufacturing semi-structured knowledge, which is the fusion of (i) general information, (ii) solution approach information, (iii) the knowledge coming from knowledgeable workers, and (iv) feedback knowledge.

General Information (I_G)

General information is described as fundamental information corresponding to a problematic event, for example, a description, observer, location and time etc. The contents are presented as follows (Fig. 11.3). The problematic event (E) is derived from event transformation function (f_E) which comprises two parameters: o: busi-

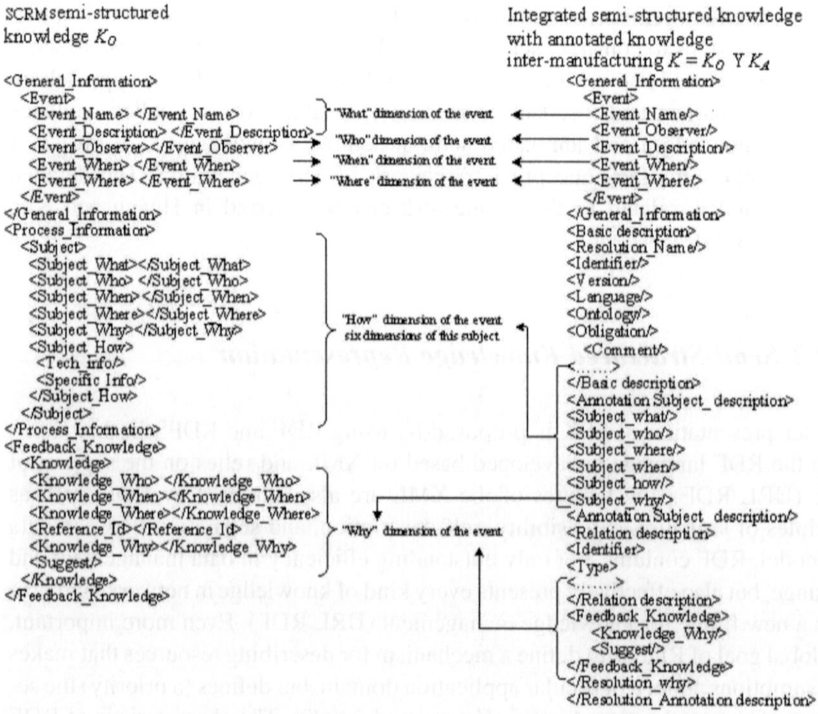

Fig. 11.2 An example of XML-based SSK documentation

```
<General_Information>
  <Event>
    <Event_Name> </Event_Name>
    <Event_Observer></Event_Observer>
    <Event_Description> </Event_Description>
    <Event_When> </Event_When>
    <Event_Where> </Event_Where>
  </Event>
</General_Information>
```

$$E = f_E(o, p)$$

Fig. 11.3 Format of XML for general information

ness entity (organization); and p: business process. Semi-structured knowledge is initiated from problematic events. The purpose of event classification is to support the distinction of semi-structured knowledge.

The problematic event is clarified and the types of structure and processes required to solve the problem are also specified. The General Information is translated to RDF and RDFS format. I_G example in RDFS format is presented in Fig. 11.4.

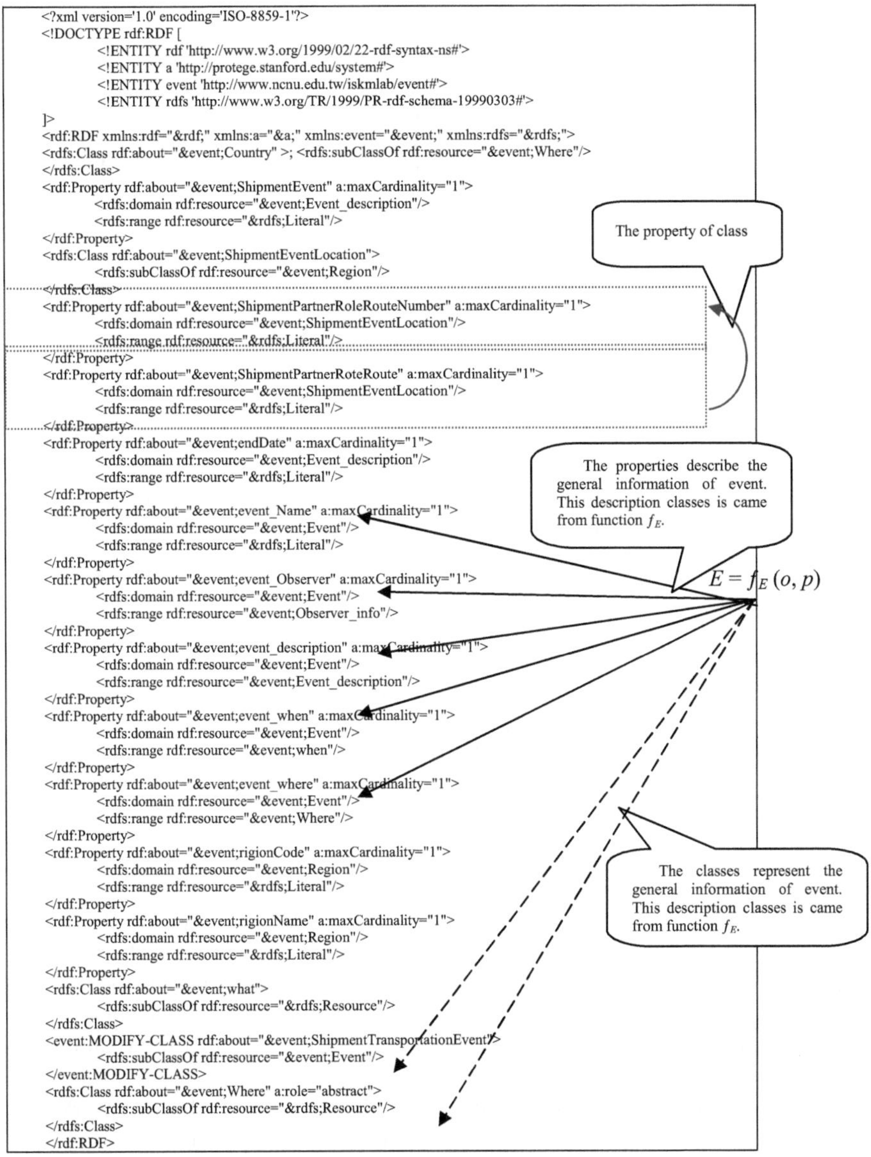

Fig. 11.4 Format of RDFS for partial general information

Solution approach Information (I_S)

The Solution approach information includes important parameters used in problem-solving methods and results. This information allows users to understand messages of solution-related subject, for example, how the solution method is to be applied. The problem-solving processes and results may not be the same due to the differ-

```
<Process_Information>
  <Subject>
    <Subject_What></Subject_What>
    <Subject_Who> </Subject_Who>
    <Subject_When></Subject_When>
    <Subject_Where></Subject_Where>
    <Subject_Why></Subject_Why>
    <Subject_How>
      <Tech_info/>
      <Specific_Info/>
    </Subject_How>
  </Subject>
</Process_Information>
```

$$S = f_s(w_1, w_2, w_3, w_4, w_5, h_1)$$

Fig. 11.5 Format of XML for solution approach information

ent methods that are used. The contents of Fig. 11.5 show important components of solution approach information, 5W1H (What, Where, Who, When, Why, and How). Where S is the subject of problematic event and f_S is the subject transformation function. The subject S should be clear, concise, and able to represent the characteristics of semi-structured knowledge. The structure of the subject should be dynamic, for example, sensitive to dimension and level changes. The entity of the subject is "what", ω_1. This parameter is most critical and dominates the rest of parameters (where, who, when, why, and how, i. e., place (ω_2), people of individual (ω_3), timing (ω_4), motivation (ω_5), and solution process (h_1)). Note that different solution approaches contain different data. The 5W1H (What, Where, Who, When, Why and How) is used to specify "Subject." The following example of RDF and RDFS format of I_S illustrates association analysis (Fig. 11.6).

Knowledge (K_P, K_T)

Knowledge (K_P, K_T), where K_P represents knowledge from the knowledgeable worker or analyzer and K_T represents feedback knowledge from the domain knowledgeable worker or administrator that includes (i) the explanation from the problem solvers for the results of solution approach information, and (ii) the feedback from other knowledgeable workers (including problem solver) to this information and knowledge.

K_P corresponds to the knowledge resulted from a knowledgeable worker or analyzer using a particular solution approach, e. g., the Apriori association rules (Agrawal and Srikant 1994). This knowledge is more focused on solution-related expertise; i. e., the knowledge is generated based on a subject S corresponding to a problematic event. K_P is derived from knowledge process function f_p which is composed of three parts: E: problematic event; S: subject, and P: knowledgeable worker (problem solver).

K_T corresponds to the feedback knowledge that provides different professional knowledge from different standpoints and perspectives of knowledgeable workers. K_T is derived from knowledge transformation function f_T which is composed of the

```
<?xml version='1.0' encoding='ISO-8859-1'?>
DTD declaration
<rdf:RDF xmlns:rdf="&rdf;" xmlns:a="&a;" xmlns:subject="&subject;" xmlns:rdfs="&rdfs;">
..........
<rdf:Property rdf:about="&subject;regionName" a:maxCardinality="1">
        <rdfs:domain rdf:resource="&subject;Region"/>
        <rdfs:range rdf:resource="&rdfs;Literal"/>
</rdf:Property>
<rdfs:Class rdf:about="&subject;when">
        <rdfs:subClassOf rdf:resource="&rdfs;Resource"/>
</rdfs:Class>
<rdfs:Class rdf:about="&subject;who">
        <rdfs:subClassOf rdf:resource="&rdfs;Resource"/>
</rdfs:Class>
<rdfs:Class rdf:about="&subject;Where">
        <rdfs:subClassOf rdf:resource="&rdfs;Resource"/>
</rdfs:Class>
<rdfs:Class rdf:about="&subject;what">
        <rdfs:subClassOf rdf:resource="&rdfs;Resource"/>
</rdfs:Class>
<rdfs:Class rdf:about="&subject;Employee_Info">
        <rdfs:subClassOf rdf:resource="&subject;Human resource_Info"/>
</rdfs:Class>
<rdfs:Class rdf:about="&subject;Expert_Info">
        <rdfs:subClassOf rdf:resource="&subject;Employee_Info"/>
</rdfs:Class>
<rdfs:Class rdf:about="&subject;Why">
        <rdfs:subClassOf rdf:resource="&rdfs;Resource"/>
</rdfs:Class>
<rdfs:Class rdf:about="&subject;How">
        <rdfs:subClassOf rdf:resource="&rdfs;Resource"/>
</rdfs:Class>
.......
.....
<rdfs:Class rdf:about="&subject;Subject">
        <rdfs:subClassOf rdf:resource="&subject;Process Information"/>
</rdfs:Class>
<rdf:Property rdf:about="&subject;Subject_How" a:maxCardinality="1">
        <rdfs:range rdf:resource="&subject;How"/>
        <rdfs:domain rdf:resource="&subject;Subject"/>
</rdf:Property>
<rdf:Property rdf:about="&subject;Subject_What" a:maxCardinality="1">
        <rdfs:domain rdf:resource="&subject;Subject"/>
        <rdfs:range rdf:resource="&subject;what"/>
</rdf:Property>
<rdf:Property rdf:about="&subject;Subject_When" a:maxCardinality="1">
        <rdfs:domain rdf:resource="&subject;Subject"/>
        <rdfs:range rdf:resource="&subject;when"/>
</rdf:Property>
<rdf:Property rdf:about="&subject;Subject_Where" a:maxCardinality="1">
        <rdfs:domain rdf:resource="&subject;Subject"/>
        <rdfs:range rdf:resource="&subject;Where"/>
</rdf:Property>
<rdf:Property rdf:about="&subject;Subject_Who" a:maxCardinality="1">
        <rdfs:domain rdf:resource="&subject;Subject"/>
        <rdfs:range rdf:resource="&subject;who"/>
</rdf:Property>
..........
</rdf:RDF>
```

The 5W1H classes which represents the general information of solution information I_S.

$S = f_S(\omega_1, \omega_2, \omega_3, \omega_4, \omega_5, h_1)$

These description properties came from function f_S which help defines the 5W1H to descript solution approach subject.

Fig. 11.6 Format of RDFS for partial solution approach information

following: K_P: knowledge from the knowledgeable worker; P': people in general (not limited to knowledgeable worker); and C_i: the ith content category.

The format of XML, which includes illustration of association analysis, is shown in Fig. 11.7.

There are two types of knowledge: knowledgeable workers' explanation for the results of solution approach information and the feedback of other knowledgeable workers for the information and knowledge. The RDFS format of basic format of

```
<Feedback_Knowledge>
  <Knowledge>
    <Knowledge_Who>
    <Id/>
    <depart/>
    <Name/>
    <Position/>
    </Knowledge_Who>
    <Knowledge_When></Knowledge_When>
    <Knowledge_Where></Knowledge_Where>
    <Reference_Id></Reference_Id>
    <Knowledge_Why></Knowledge_Why>
    <Suggest/>
  </Knowledge>
</Feedback_Knowledge>
```

$$K_P = f_P(E,S,P)$$
$$K_T = f_T(K_P,P',C_i)$$

Fig. 11.7 Format of XML of feedback knowledge

$K_P + K_T$ are generated. The generating of semi-structured knowledge represented within the Zachman Framework clarifies the processes and activities of knowledge generation in the organization. This semi-structured knowledge formulated as RDF and RDFS documents are helpful in capturing, storing, managing, and sharing knowledge in organizations over the semantic web.

11.3 Annotation Framework for Supply Chain Risk Management Knowledge

The annotation processes is shown in Fig. 11.8. The SCRM knowledge documents in heterogeneous format are solved and stored in the following three steps: (Heflin and Hendler 2000):

Step 1: Each SCRM knowledge contents are captured from un-structured or heterogeneous SCRM knowledge documents. Attach XML-description tags on SCRM knowledge contents. Then the XML-based metadata for this SCRM knowledge content is generation.

Step 2: Meta-datum of knowledge contents are classified into basic description information (D_B), annotation description information (D_A), relation description information (D_R) and transformed into the documents in RDF and RDFS format. The annotated SCRM knowledge (K_A) in SSK model is generated. Conceptual schemas are also created in this step. The conceptual schema is a RDFS file which is used to describe structure of the concepts and properties in the annotated SCRM knowledge. This schema is used in querying required information of concept by mapping onto the benchmark ontology. Here, it is presumed that the benchmark ontology can offer sufficient information (e. g. instances, constraints, and relationships) for these concepts (Theobald and Weikum, 2002).

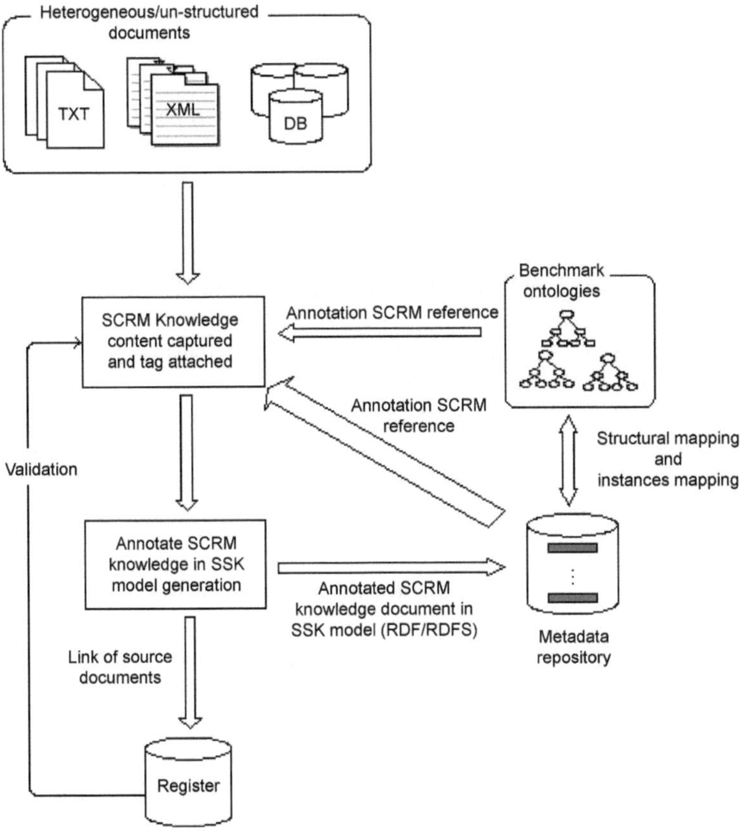

Fig. 11.8 Annotation processes

Step 3: (1) Store annotated SCRM knowledge document with conceptual schema into the metadata repository.
(2) Store the mapping results (e. g., metadata) between the benchmark ontology and the annotated SCRM document into the metadata repository.
(3) Register the link to the original SCRM document in the register.

The metadata repository is capable to validate the availability of metadata and the consistence between the elements in the benchmark ontology and the metadata of the original sources SCRM documents. The register validates the link consistence between the annotated SCRM knowledge document and original SCRM knowledge source.

In practice, industries should recognize the changes in the SCRM environment. Particularly, each change in the SCRM conceptualization, for example, changes in the strategy of the company, in the production planning, and in the inventory segmentation, that requires an update of ontology. The requirement of the ontology update is often discovered in extracting and mapping the metadata into the benchmark ontology. The annotation agent should update the benchmark ontology when the requirement is fulfilled.

The SCRM annotation consists of three parts of description, namely basic description (D_B), annotation description (D_A), and relationship description (D_R).

- Basic description (D_B) illustrates the general statements of the SCRM knowledge document and its source.
- Annotation description (D_A) is based on the 5W1H Zachman framework (What, Where, Who, When, Why, and How) to represent annotated SCRM knowledge in the SSK model.
- Relationship description (D_R) illustrates two types of relationship: (i) the relationship between the annotated SCRM knowledge document and other heterogeneous/unstructured SCRM documents and (ii) the relationship between the benchmark ontology and heterogeneous ontology. For example, while SCRM documents are clustered into several groups, the meta-data in each SCRM document in the same group records both its own document id and id(s) of other documents in the same group.

Figure 11.9 illustrates the annotation format using XML tag instead of RDF and RDFS. The notation B, in Fig. 11.9, represents the basic description of annotated subject, used as fundamental information, corresponding to a subject that needs to be annotated. For example, name, identifier, registry, version and language SCRM. B is derived from the basic description function f_B which is composed of the following: I_D: solution-related subject document identification; and O_D: SCRM ontology identification. B corresponds to the relationship among (i) SCRM knowledge document identification, (ii) ontology information that the SCRM knowledge responds to, and (iii) a basic description of annotated SCRM knowledge.

The notation R (relation description of annotated subject) in Fig. 11.9 describes the relation information between annotated SCRM knowledge documents and their related knowledge sources (e. g. other organization's ontology). R is derived from relation indication function f_R which is composed of the following: I_R: relationship information; and t: the relation type. R is a group of source tags to describe (i) the relationship between the annotated SCRM knowledge document, other heterogeneous SCRM documents, (ii) the relationship between the benchmarking ontology, and heterogeneous ontology. Note that SCRM annotation description does not just include one level 5W1H (What, Where, Who, When, Why, and How) structure. Additional levels of 5W1H description tag are dynamically structured and used for more detailed description.

Illustrative example: A company posts a problem to search for the solution in SCRM: The defective rate of the Printed Circuit Board (PCB) installed in the electronic device is relatively high in this month. After thoroughly investigating the case, it appears the issue is about faulty chip insertion during the assembly stage. To reduce the defective PCB and the risk of supply chain shutting down and avoid the delay of producing the final product, it is suggested that launching an advanced Radio Frequency ID (RFID) system to improve the chip ID recognition. The solu-

<Basic description>
<Name/> - The label assigned to the SCRM metadata subject
<Identifier/> - The unique identifier assigned to the SCRM metadata subject
<Version/> - The version of the SCRM metadata subject
<Language/> - The language in which the SCRM metadata subject is specified
<Ontology/> - A statement that clearly represents the concept and essential nature of the SCRM metadata
<Obligation/> - Indicates if the SCRM metadata subject is required to always or sometimes be present
<Comment/> - A remark concerning the application of the SCRM metadata subject
<.......>
</Basic description>

$B = f_{B'}(I_D, O_D)$

<Annotation description>
<what/>- contents of the subject to activity
<who/>- something involved in subject and activity
<where/>- Locations of the subject
<when/>- Time of the subject
<why/>- motivations of subjects and reasons of activity
<how/>- Solution approaches of the subject
</Annotation description>

$A = f_A(w_1, w_2, w_3, w_4, w_5, h_1)$

<Relation description>
<Identifier> - The related unique identifier to the source
<Type>-The relation type
<Art_info>articulation info (if articulation process trigger)
<........>
</Relation description>

$R = f_R(I_R, t_i)$

Fig. 11.9 SCRM annotation format presented in XML tag

tion is responded in HTML format and the entire interaction process is presented in a knowledge representation enhancement system (KRES). In general, this type of support system is suitable to use in computer industry to augment inventory/logistics control. The details of system operations are as follows:

First, the solution-related document is received and browsed with Internet Explorer (IE) (Fig. 11.10).

Through the annotation process, the SCRM document is transformed into semi-structured format. In this step, the subject information is obtained from the source document and classified into three description categories of description, (D_B, D_A, D_R), through the semi-automatic analysis and validation operated by human annotation expert or annotation agents. An annotation agent or annotation expert (i) parses the source HTML document, and knowledge contents required to be annotated are captured; (ii) analyzes the semantic hiding in each knowledge content, and searches the related meta-datum in the metadata repository as well as the suitable elements in the benchmark ontology; (iii) bases on the meta-datum and the elements to generate XML-based description tags to annotate the SCRM knowledge content; (iv) if the search reaches nothing, the annotation agent and expert infers and identifies the

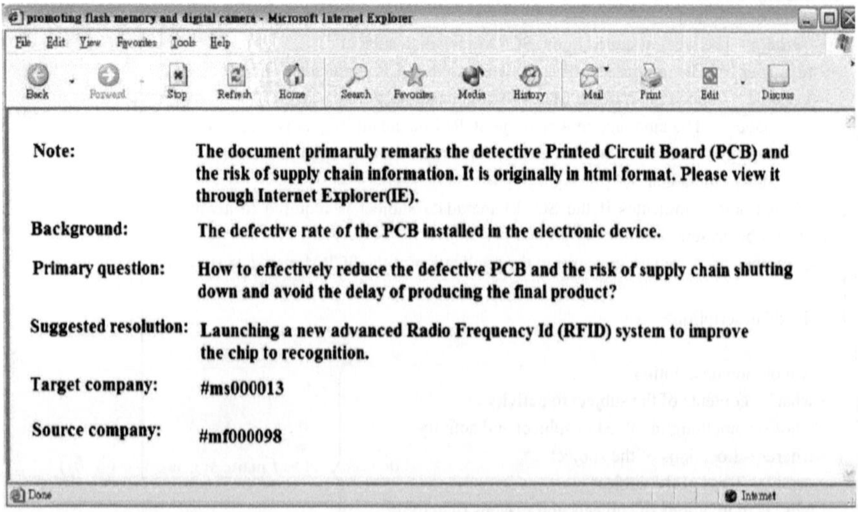

Fig. 11.10 The original HTML document before annotated browsed in IE

semantic hiding in the knowledge contents, then creates a new XML-based description tag to annotated the SCRM knowledge content; (v) classifies SCRM knowledge content with description tag into the adequate categories, (D_B, D_A, D_R). The annotated SCRM knowledge of the original document in XML-based Semi-Structured format is generated. For example, D_B, basic description (Name, Identifier, Ontology, Obligation, and Comment) is captured and annotated on top rows of the source document (Document_Id, Note, and primary question). Moreover, the message on the row "Note" indicates that the knowledge belongs to the defective Printed Circuit Board (PCB) and the risk of supply chain ontology. Therefore, the information "the defective PCB" and "risk of supply chain" are attached to the tag "ontology," which also allows other knowledge user realize these SCRM knowledge document is related to the defective Printed Circuit Board (PCB) and the risk of supply chain ontology. The knowledge contents represented in text format in rows "Primary Question" and "Suggested resolution" of Fig. 11.10, are also captured and annotated into 5W1H (Zachman framework), where knowledge content in "Primary Question" corresponds to the "What", "When", "Where", "Who", and "Why" dimensions, respectively, and knowledge content in "Suggested resolution" corresponds to the "How" dimension and annotated with "Sales Promotion" tag. The generated XML in SSK model is presented as follows (Fig. 11.11).

Next, based on the XML-based SSK model, the annotated semi-structured knowledge (K_A) and the conceptual schemas are generated and stored in RDF and RDFS, respectively. In this step, the annotation agent who is responsible to (i) transforms the XML-based description tags into concepts and properties; and SCRM knowledge contents into instances and values (ii) constructs the conception schema in RDFS for the structure of annotated SCRM knowledge (e. g. concepts hierarchy,

```
<Basic description>
  <Name> SCRM_q1.html</Name>
  <Identifier>\\IASERVER\SCRM\MA000001025</Identifier>
  <Version> 1.0</Version>
  <Language> English</Language>
  <Ontology> the defective PCB ; Risk of supply chain </Ontology>
  <Obligation>department vision</Obligation>
  <Comment>The source was created in HTML</Comment>
  ………
</Basic description>

<Annotation description>
  <who><entry> RFID tag</entry><entry >The upper stream supplier s </entry></who>
  <where><country>Taiwan</country></where>
  <when><Yesr>2008</Year><Quarter>Q2</Quarter></when>
  <what>< formula> To attach the RFID tag</formula></what>
<why><Reason> Faulty chip insertion during the assembly stage.</Reason></why>
  <how>< formula> To attach the RFID tag in the chip between the supplier and the manu-
facturer </formula></how>
  </Annotation description>

<Relation description>
  <Identifier>\\IASERVER\SCRM\MA000000852</Identifier>
  <Type>Cluster</Type>
  ……
</Relation description>
```

Fig. 11.11 (D_B, D_A, D_R) represented with XML

and concepts class and its corresponding properties); (iii) annotates these concepts and their instances based on the RDF syntax. Last, a RDF and RDFS document describes the annotated SCRM knowledge in SSK is completed. The exemplified annotated SCRM knowledge content in RDF is presented in Fig. 11.12.

The conception schema of the exemplified annotated SCRM knowledge document is partially described in Fig. 11.13. Through the annotation process, the formal and consistent meta-datum from the annotated SCRM knowledge document in SSK model is generated. By viewing the annotated SCRM knowledge in SSK model, knowledgeable worker can easily realize the purpose, contents of the heterogeneous and unstructured SCRM knowledge document. Furthermore, the annotations are based on RDF standard; therefore, metadata is machine understandable on the Semantic Web. In this annotated SCRM knowledge document, every resource in RDF and RDFS is annotated an URI (Uniform Resource Identifiers), which makes the annotated SCRM knowledge sources reusable and machine processable on the web (Decker, 2002).

```
<?xml version='1.0' encoding='ISO-8859-1'?>
<!DOCTYPE rdf:RDF [
        <!ENTITY rdf 'http://www.w3.org/1999/02/22-rdf-syntax-ns#'>
        <!ENTITY AnnotatedSSK_mo 'http://www.ncnu.edu.tw/im/iskmlab/AnnotatedSSK_model#'>
        <!ENTITY rdfs 'http://www.w3.org/TR/1999/PR-rdf-schema-19990303#'>
]>
<rdf:RDF xmlns:rdf="&rdf;" xmlns:AnnotatedSSK_mo="&AnnotatedSSK_mo;" xmlns:rdfs="&rdfs;">
<AnnotatedSSK_mo:quarter rdf:about="&AnnotatedSSK_mo;AnnotatedSSK_00079"
        AnnotatedSSK_mo:Quarter="Q2"
        AnnotatedSSK_mo:Year="2008"/>
<AnnotatedSSK_mo:Country rdf:about="&AnnotatedSSK_mo;AnnotatedSSK_00080"
        AnnotatedSSK_mo:Countryname="Taiwan"/>
<AnnotatedSSK_mo:What rdf:about="&AnnotatedSSK_mo;AnnotatedSSK_00068"
        AnnotatedSSK_mo:formual=" To attach the RFID tag "/>
<AnnotatedSSK_mo:Who rdf:about="&AnnotatedSSK_mo;AnnotatedSSK_00071"
        AnnotatedSSK_mo:entry="RFID tag "/>
<AnnotatedSSK_mo:Who rdf:about="&AnnotatedSSK_mo;AnnotatedSSK_00072"
        AnnotatedSSK_mo:entry=" The supplier and the manufacturer "/>
<rdf:Description rdf:about="&AnnotatedSSK_mo;AnnotatedSSK_00046"
        AnnotatedSSK_mo:ADid="AD1">
        <AnnotatedSSK_mo:when rdf:resource="&AnnotatedSSK_mo;AnnotatedSSK_00079"/>
        <AnnotatedSSK_mo:where rdf:resource="&AnnotatedSSK_mo;AnnotatedSSK_00080"/>
        <rdf:type rdf:resource="&AnnotatedSSK_mo;Annotation description"/>
        <AnnotatedSSK_mo:what rdf:resource="&AnnotatedSSK_mo;AnnotatedSSK_00068"/>
        <AnnotatedSSK_mo:who rdf:resource="&AnnotatedSSK_mo;AnnotatedSSK_00072"/>
        <AnnotatedSSK_mo:why rdf:resource="&AnnotatedSSK_mo;AnnotatedSSK_00057"/>
        <AnnotatedSSK_mo:how rdf:resource="&AnnotatedSSK_mo;AnnotatedSSK_00058"/>
</rdf:Description>
<AnnotatedSSK_mo:Why rdf:about="&AnnotatedSSK_mo;AnnotatedSSK_00057"
        AnnotatedSSK_mo: Reason =" Faulty chip insertion during the assembly stage "/>
……..
</AnnotatedSSK_mo:How>
<rdf:Description rdf:about="&AnnotatedSSK_mo;AnnotatedSSK_00060"
        AnnotatedSSK_mo:Comment="SCRM_q2.html"
        AnnotatedSSK_mo:Identifier="1.0"
        AnnotatedSSK_mo:Language=" English "
        AnnotatedSSK_mo:Name="The source was created in HTML"
        AnnotatedSSK_mo:Obligation="\\IASERVER\Marketing\MA000001025"
        AnnotatedSSK_mo:Version="department vision">
        <rdf:type rdf:resource="&AnnotatedSSK_mo;Basic description"/>
        <AnnotatedSSK_mo:Ontology> the defective PCB </AnnotatedSSK_mo:Ontology>
        <AnnotatedSSK_mo:Ontology>Risk of supply chain</AnnotatedSSK_mo:Ontology>
</rdf:Description>
<rdf:Description rdf:about="&AnnotatedSSK_mo;AnnotatedSSK_00061"
        ……..
</rdf:RDF>
```

"5W1H" Annotation of Knowledge Subject Event

"Basic Annotation" of Knowledge Subject Event, where basic identifier and related ontology

Fig. 11.12 Exemplified annotated SCRM knowledge document stored in RDF format

```
    <rdf:Property rdf:about="&AnnotatedSSK_mo;ADid" a:maxCardinality="1" >
        <rdfs:domain    rdf:resource="&AnnotatedSSK_mo;Annotation    descrip-
tion"/>
        <rdfs:range rdf:resource="&rdfs;Literal"/>
    </rdf:Property>
    <rdfs:Class rdf:about="&AnnotatedSSK_mo;Annotation description"/>
        <rdfs:subClassOf rdf:resource="&rdfs;Resource"/>
    </rdfs:Class>
    <rdfs:Class rdf:about="&AnnotatedSSK_mo;Basic description">
        <rdfs:subClassOf rdf:resource="&rdfs;Resource"/>
    </rdfs:Class>
    <rdf:Property rdf:about="&AnnotatedSSK_mo;Comment"
         a:maxCardinality="1">
        <rdfs:domain rdf:resource="&AnnotatedSSK_mo;Basic description"/>
        <rdfs:range rdf:resource="&rdfs;Literal"/>
    </rdf:Property>
    <rdfs:Class rdf:about="&AnnotatedSSK_mo;Country">
        <rdfs:subClassOf rdf:resource="&AnnotatedSSK_mo;Where"/>
    </rdfs:Class>
    <rdf:Property                   rdf:about="&AnnotatedSSK_mo;Countryname"
:maxCardinality="1">
        <rdfs:domain rdf:resource="&AnnotatedSSK_mo;Country"/>
        <rdfs:range rdf:resource="&rdfs;Literal"/>
    </rdf:Property>
    ..........
```

The concepts, properties and their hierarchy relationships is identified in RDFS of the annotated document

Fig. 11.13 Partial conceptual schema of the exemplified annotated SCRM knowledge document

11.4 Interaction of Annotated SCRM Knowledge Documents with the Benchmark Ontology

The knowledge shared in a SC has been presented in a flexible and well-format knowledge representation structure. In general, industries generate different documents in SSK model for different purposes. In order to avoid the redundancy of the use of meta-datum, an approach to coordinate and manage meta-datum in the metadata repository and benchmark ontology is required. After organizations receiving these annotated SCRM documents, the important step, extracting and mapping the meta-datum of annotated SCRM knowledge document into benchmark ontology is processed.

11.4.1 Extracting and Mapping

In addition to annotating SCRM knowledge documents into SSK model, the annotation agent further performs the role of extractor (Schenk, 1999) that retrieves the string-values of all elements in the annotated semi-structured SCRM knowledge and performs pertinent functions on them in the following methods: (i) decomposing the meta-datum (annotated knowledge elements) of annotated SCRM documents into basic elements; (ii) extracting these basic elements and their corresponding URI. Finally, the meta-datum are extracted and used to map them to the corresponding elements of the benchmark ontology.

Figure 11.14 presents the architecture of extraction. The basis for the development of the application program is the Document Object Model (DOM) application interface (API) (URL DOM), a recommendation of the W3C for accessing and modifying HTML and XML-based documents. A streaming API for accessing the SCRM documents is not suitable, since the annotation agent at extracting step may contain forward and backward reference to find elements in SCRM documents. These references are required to be resolved in real time such that the complete SCRM document in SSK model can be accessed in time. (Huck et al., 1999)

Mapping is a process that bridges, transforms, and unifies knowledge elements of semi-structured knowledge into the target equivalence or relative elements of the benchmark ontology.

Mapping is an operator (Frasincar et al., 2002) defined as follows:

$$\cup(S(e_1),\ S(e_2),\ \ldots,\ S(e_n)),$$

where ∪ corresponding to a collection, S to a set, and $e_1, \ldots, e_n$ to elements.

The elements are concepts, classes, or properties as well as instances of these concepts classes in both SCRM knowledge document and the benchmark ontology.

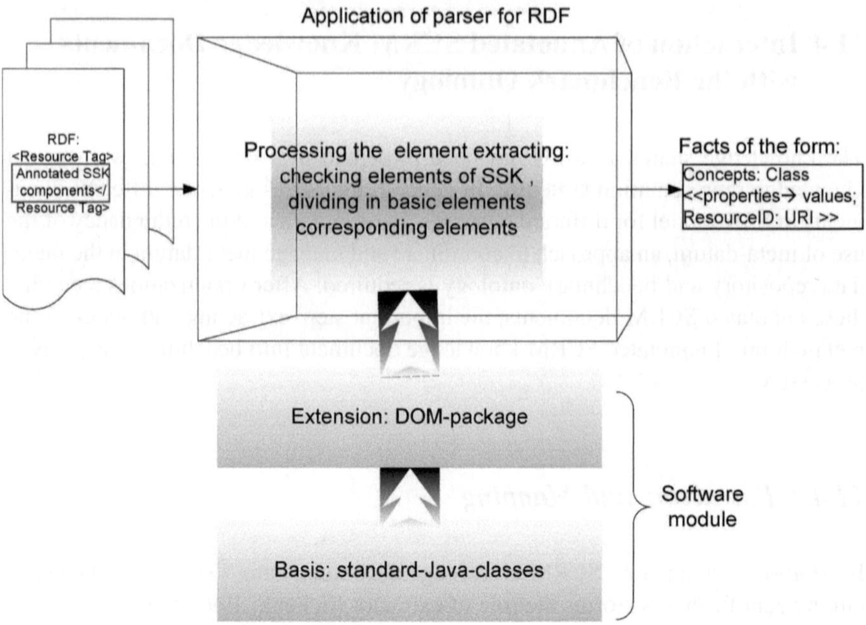

Fig. 11.14 Architecture of the Extraction

The mapping operator combines (set union) elements in the SCRM documents in SSK model and the benchmark ontology to obtain the final mapping result. In the annotation process, by storing the mapping results and together with the annotated SCRM document into the metadata repository, a degree of local autonomy of annotated SCRM knowledge document to coexist with content consistency (Fensel et al., 2000) of partial interoperability of the benchmark ontology could be achieved.

Furthermore, by using extracting and mapping, the knowledge elements could not only be retrieved from annotated SCRM knowledge documents and mapped to the benchmark ontology, but also be composed "from" the benchmark ontology vice versa. Figure 11.16 illustrates the two-way interaction between the benchmark ontology and SSK. That is, organizations can generate SCRM knowledge documents in SSK model by constructing from the benchmark ontology concepts, properties and its values of instances. Note that the benchmark ontology of supply chain illustrated in Fig. 11.15 is referenced to Chen (2003).

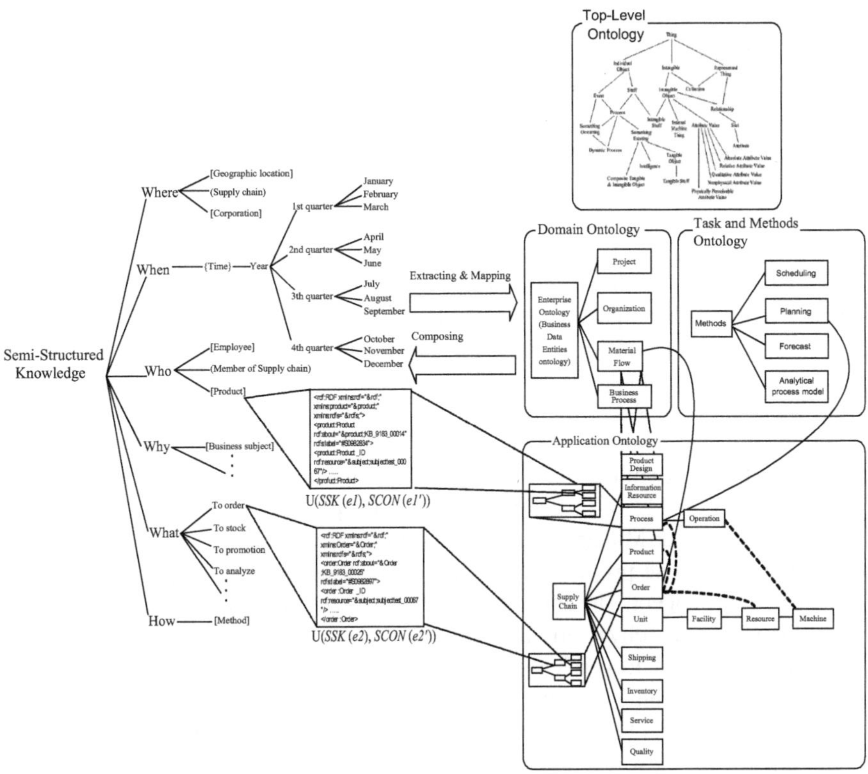

Fig. 11.15 The interaction between SSK and the benchmark ontology

11.4.2 Interaction Between SSK and the Benchmark Ontology

The main purpose of the interaction is to extract concepts, properties, instances from the SCRM documents in SSK model, compare them with the elements of benchmark ontology, and induce the extracted elements in desired taxonomies to support the elements of benchmark ontology. Two types of interaction are:

(1) Structural interaction: extracting concepts and properties of semi-structured SCRM knowledge documents in order to map them into the benchmark ontology.
(2) Instance-based interaction: extracting instances of semi-structured SCRM knowledge documents in order to for map them into benchmark ontology.

Structural interaction: Extracting concepts and properties from the conceptual schema of annotated SCRM knowledge document, and map these concepts and properties into the corresponding concepts and properties in the benchmark ontology.

In this semi-structured knowledge, concepts have its hierarchical relationship corresponding to the knowledge application domain. For example, grey area in Fig. 11.12, the hierarchical structure of concepts represented with the "where" dimension is corresponding to the "location" concepts in the benchmark ontology, where the manufacturing process holds. Furthermore, the contents of location are represented by "country" and "country name" respectively. In Fig. 11.12, "Taiwan" is used as a country name.

The annotation agent analyzes and extracts (i) concepts and (ii) properties from RDFS of annotated SCRM knowledge documents (conceptual schema). Then the agent searches and compares them with all concepts and properties in the benchmark ontology. If the concept (or property) is equivalent or related to extracted concepts (or properties), the mapping between the two concepts (or properties) is bridged from the conceptual RDFS file to benchmark ontology.

Following the mapping example of Fig. 11.12, the sub-concept and properties of the concept corresponding in the "where" dimension are presented in the left side paragraph of Fig. 11.16. In the paragraph, concept "where" has one sub-concept "Supply Chain Risk Management". Within the "Supply Chain Risk Management" concept, there is one property "Suppliername" describes the "Supply Chain Risk Management" concept. The annotation agent analyzes these concepts and properties in a three-level hierarchy relationship, searches and compares related concepts in the benchmark ontology to verify the mapping. Finally, the agent finds a concept "the defective PCB Planning" which is subclass of "Supply Chain Risk Management" is matched to concepts of SSK document. Then the annotation agent maps the concept "Supply Chain Risk Management" to the concept "the defective PCB Control" in the benchmark ontology (on the right side of Fig. 11.16).

The map on the concepts is described as follows:

$\cup$ (AnnotatedSSK(where. Supply Chain Risk Management, COntology.Enterprise (the defective PCB Control)) ... (Mapping 1),

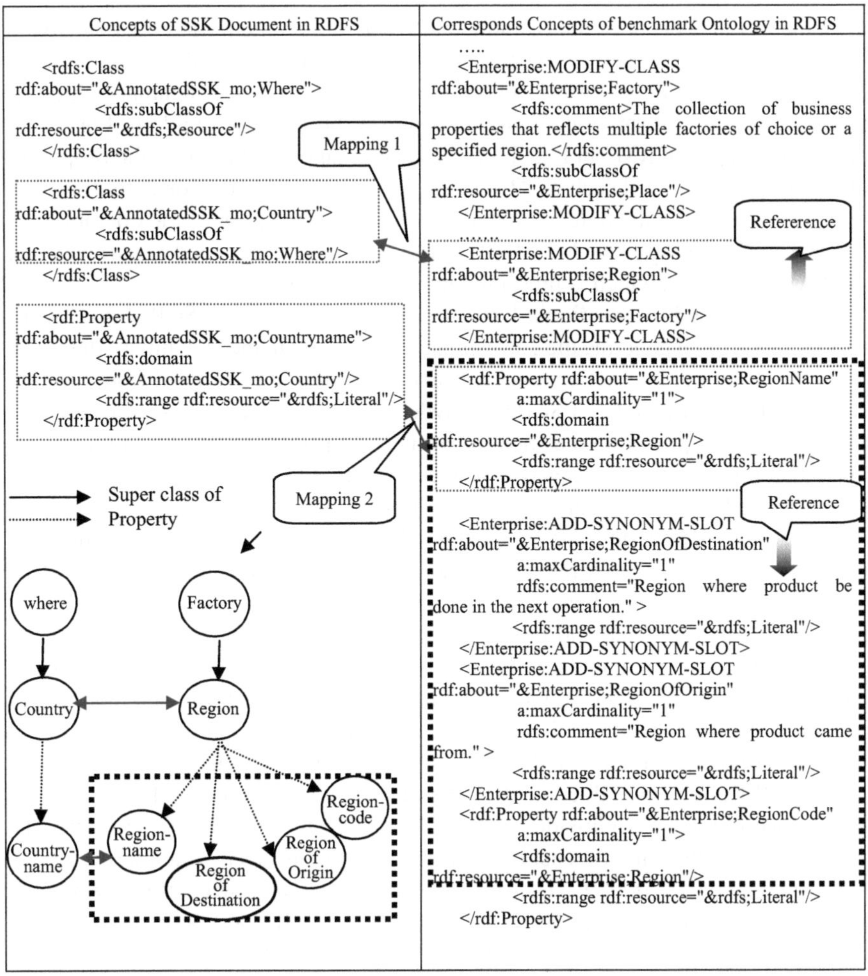

Fig. 11.16 The mapping between the benchmark ontology and the conceptual schema

Mapping 1 is indicated with the two-way solid arrowhead line in Fig. 11.16. ∪ (AnnotatedSSK(what. Supply Chain Risk Management, (Suppliername)), *SCOntology.Enterprise*(Tool … *the defective PCB Control*)).(*ToolName*))) … (Mapping 2).

Mapping 2 indicates that the property "*Suppliername*" in the annotated document is mapped into the property "ToolName" in benchmark ontology (a double heads arrow with solid line in Fig. 11.16). Replace country name with type of supply chain risk management.

These mapping relationships are stored together with the annotated SSK document in the metadata repository. By utilizing the mapping, other application agents and knowledgeable workers, besides the annotation agent, can easily recognize the meaning of concepts between the annotated SCRM documents and the benchmark ontology. The agents and workers can track concepts in the SCRM documents, through the mapping backward to corresponding concepts in benchmark ontology and obtain more information about their parents, brothers, or children. For example, the concept "SCRM" in Fig. 11.16 has been mapped. Therefore, the properties of the inventory control such as "ToolName"(the name of operation activities; i. e., replenishment etc), "the defective PCB control code"(the code of the operation activities), "PSBOfOrign", and "PCBOfDestination" can be extended to retrieve more SCRM information, which is presented in dot line block.

Instance-based interactions: Extract instances of annotated SCRM documents in SSK model then generate the corresponding instances in the benchmark ontology.

In the structural interaction, concepts and properties are extracted from the conceptual schema of annotated SCRM knowledge document, and these concepts and properties are mapped into the corresponding concepts and properties in the benchmark ontology. In the same way, instances of these concepts and properties are also extracted and mapped in the instance-based interaction. The main purpose of instance-based interaction allows the knowledgeable workers to obtain more instances by mapping the extracted instance onto the corresponding instance (e. g., belong to property A) in benchmark ontology, and the other instances (belong to property A) are extracted and presented to the knowledgeable workers. All of instances from a document in the SSK map themselves into the ones in the benchmark ontology and obtains more instances for knowledgeable workers. The mapping of instances just likes "recording" the instances into benchmark ontology.

The example in Fig. 11.17 illustrates the instance mapping between the annotated SCRM knowledge document in Fig. 11.12 (left-side in Fig. 11.17) and the benchmark ontology (right-side in Fig. 11.17). In the left-side of the figure, the concepts in the "when" dimension is assigned to the instance which has the URI: "mo;AnnotatedSSK_00079." That is, the instance has the URI: http://www.ncnu.edu.tw/im/iskmlab/AnnotatedSSK_model#AnnotatedSSK_00079, which indicates the concept "quarter" has one instance that has property Quarter="Q2" and Year="2008". The concept class of "Date Period" in the right-side of Fig. 11.16 generates the corresponding instance which has the URI: "&Enterprise;SCDomainOntology_00567", where the property Year="2008" , beginDate="2008/04", endDate="2008/06", quarter="Q2".

The instance mapping corresponding to the solid two-way arrowhead line in Fig. 11.17 is formatted as follows:

∪ *instance*(*AnnotatedSSK*(*when*(*quater*):*AnnotatedSSK_00079*), *SContology.Enterprise*(*Time_info.*(*DatePeriod*): *SCDomainOntology_000567*)) . . . (Mapping 3) .

Mapping 3 shows that the instance "annotatedSSK_00079" in the annotated SCRM document corresponds to the instance "*Enterprise*;SCDomainOntology_00567" in the benchmark ontology.

The same as above, the other annotating instances of the concept in the annotation document are mapped to their corresponding instances in the benchmark ontology.

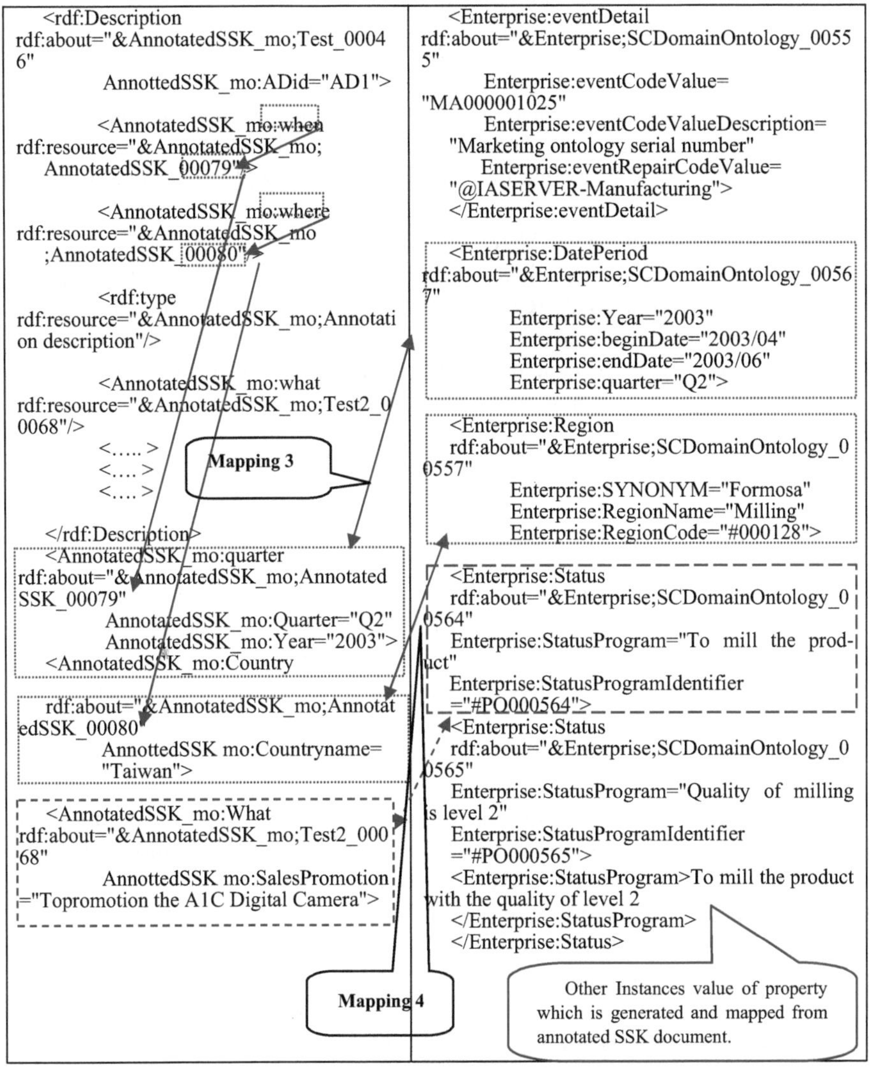

Fig. 11.17 The instance mapping between the example (Figure 12) and benchmark ontology

For example,
$\cup$ *instance(AnnotatedSSK(where SCRM(countryname):AnnotatedSSK)_00080, SCOntology.Enterprise(Place.Factory.Region.(RegionName,RegionCode):SCDomainOntology_00557)*
… (Mapping 4).

The Mapping 4 belongs to instances mapping from the concepts in the "where" dimensions of SSK document onto the instance of "Region" concept in the benchmark ontology. The property value of the "RegionName" of the mapped instances is “replenishing”, which have additional information about the mapped instances, such as SYNONYM = “Re-stock” OperationName= “replenishing”, and OperationCode= “#000128” in the benchmark ontology.

Other dimensions, e. g., “what” are also mapped and is illustrated in the block with a dash line and mapped by a double heads arrow with dash line in Fig. 11.16, where it obviously indicates that the instance of property “StatusProgram” = "To replenish the product."

The annotated documents in SSK model and the related conceptual schema (RDF Schema) are stored to the metadata repository and the links (index and location) of original source documents are registered in the register. The elements of annotated SCRM knowledge document in SSK model is extracted and mapped into the corresponding elements in the benchmark ontology. The information of heterogeneous SCRM knowledge document is captured and reused.

11.5 Conclusions

The SSK model was proposed to (i) make representation of SCRM knowledge consistent and flexible (ii) share industry SCRM knowledge with not only in meaningful, but also an explicit sharable manner over the web in organization. The SCRM knowledge documents in the SSK model, which is based on Zachman Framework (Inmon and Zachman, 1997), and the technique of knowledge representation-Resource Description Framework (RDF), were very effective to share on the semantic web.

The annotation process resolves heterogeneity of SCRM knowledge document from the various hetero-purpose information systems. The process allows organizations efficiently access and retrieve the hetero-format or unstructured SCRM knowledge documents. Moreover, it also supports the original HTML documents to be converted to the XML documents. Finally, the exemplified annotated MP knowledge documents are stored in RDF format

For different purposes, industries may generate different SCRM documents in SSK model. In order to avoid the redundancy of meta-datum, an approach was proposed to coordinate and manage meta-datum and benchmark ontology, which is called interaction of annotated MP knowledge documents with the benchmark ontology. Through the proposed approach, the organization is able to generate unified documents, which significantly improve degree of knowledge share.

The future research should emphasize (i) semantic logic inference engine for peer application agents, and (ii) the management of mapping results of annotated SCRM knowledge document. When increase amount of annotated SSK document are stored in the metadata repository the mapping results are also increased. How to manage this mapping, extract necessary information, and to aid the accuracy of annotation process deserves further research.

Acknowledgements This work was partially supported by funding from the Nation Science Council of the Republic of China. (NSC 96-2416-H-260-005-MY2; NSC-97-3114-P-260-001-Y).

References

R. Agrawal, and R. Srikant, "Fast algorithms for mining association rules in large databases", *Proc. 20th Int Conf. Very Large Data Bases*, pp. 478–499, 1994

J.L. Badaracco, "The knowledge link: How firms compete through strategic alliances", *Harvard Business School Press*, Boston, MA, 1990

M.C. Chen, "Knowledge sharing in supply chain with semantic web", *Dissertation*, Department of Information Management, National Chi-Nan University, Taiwan, 2003

Z. Cui, D. Jones, and P. O'Brien, "Issues in ontology-based information integration", *Proc. of the IJCAI-01 Workshop on Ontologies and Information Sharing,* Seattle, USA, pp. 141–146, 2001

S. Decker, "Semantic web methods for knowledge management", *Dissertation, Department of Computer Science*, University of Fridericiana zu Karlsruhe, 2002

S. Decker, M. Erdmann, D. Fensel, and R. Studer, "Ontobroker: Ontology based access to distributed and semi-structured information. Database semantics: Semantic issues in multimedia systems", R. Meersman et al. (eds.), Kluwer Academic Publisher, Boston, pp. 351–369, 1999

R. Dieng, O. Corby, A. Giboin, and M. Ribière, "Methods and tools for corporate knowledge management", *Journal of Human-Computer Studies*, 51(3), pp. 567–598, 1999

M. Erdmann, A. Maedche, H.-P. Schnurr, and S. Staab, "From manual to semiautomatic semantic annotation", *Proc. COLING 2000, Luxembourg*, 2000

D. Fensel, I. Horrocks, F. Van Harmelen, S. Decker, M. Erdmann, and M. Klein, "OIL in a nutshel", *In Knowledge Acquisition, Modeling, and Management, Proceedings of the European Knowledge Acquisition Conference (EKAW-2000),* R. Dieng et al. (eds.), *Lecture Notes in Artificial Intelligence, LNAI*, Springer, 2000

F. Frasincar, G.J. Houben, R. Vdovjak, and P. Barna, "RAL: An algebra for querying RDF", *The 3rd International Conference On Web Information Systems Engineering (WISE 2002),* T.W. Ling, U. Dayal, E. Bertino, W.K. Ng, A. Goh (eds.), Singapore, pp. 173–181, IEEE Computer Society, 2002

J. Heflin, and J. Hendler, "Searching the web with SHOE. Artificial intelligence for web search", *AAAI workshop*, WS-00-01, AAAI Press: Menlo Park, pp. 35–40, 2000

C.C. Huang and C.M. Kuo, "Transformation and searching of semi-structured knowledge in organizations", *Journal of Knowledge Management*, 7(4), pp. 106–123, 2003

G. Huck, I. Macherius, and P. Fankhauser "PDOM: lightweight persistency support for the document object model", In: *Proceedings of the 1999 OOPSLA Workshop "Java and Databases: Persistence Options"*. Held at the 14th Annual ACM SIGPLAN Conference on Object-Oriented Programming Systems, Languages, and Applications (OOPSLA'99) 1–2 November. Denver, Colorado, 1999

W.H. Inmon, J.A. Zachman, and J.G. Geiger, *Data Stores, Data Warehousing, and the Zachman Framework: Managing Enterprise Knowledge*, McGraw-Hill Companies, Inc. NY, USA, 1997

V. Kashyap and A. Sheth, "Semantic heterogeneity in global information systems: The role of metadata, context and ontologies", M. Papazoglou and G. Schlageter (Eds.): *Cooperative Information Systems: Current Trends and Directions, Academic Press,* New York, 1996

J. Li, L. Zhang, and Y. Yu, "Learning to generate semantic annotation for domain specific sentences", *knowledge markup and semantic annotation workshop in K-CAP 2001*, Dept. of Computer Science and Engineering, Shanghai JiaoTong University, Shanghai, China, 2001

D. Malone, "Knowledge management A model for organizational learning", *International Journal of Accounting Information Systems*, 3(2), pp. 111–123, 2002

P. Martin and P. Eklund, “Embedding knowledge in web documents”, *Proc. 8th Int. World Wide Web Conf. (WWW‘8)*, Toronto, pp. 1403–1419, 1999

O. Nabuco, K. Drira, and E. Dantas, “A layered design model for knowledge and information sharing cooperative systems”, *Enabling Technologies: Infrastructure for Collaborative Enterprises, WET ICE, 2001 Proceedings. The Tenth IEEE International Workshops on*, Cambridge, MA, USA, pp. 305–310, 2001

I. Nonaka and H. Takeuchi, “The Knowledge Creating Company: How Japanese Companies Create the Dynamics of Innovation”, *Oxford University Press,* New York, 1995

M. Schenk, “Ontology-based semantical annotation of XML”, *Diploma Thesis, Institute AIFB*, University of Karlsruhe, 1999

A. Theobald and G. Weikum, “The XXL search engine: Ranked retrieval of XML data using indexes and ontologies”, *Demo Paper, ACM SIGMOD International Conference on Management of Data*, Madison, Wisconsin, 2002

URL DOM: http://www.w3.org/DOM/

URL RDF: http://www.w3.org/TR/rdf /

C. Woodworth and H. Kojima, “Knowledge creation through task model and information organizing method using image/drawing expression”, *Applications and the Internet 2002 (SAINT 2002). Proceedings. 2002 Symposium on,* 46–53, Nara, Japan, 2002

P. Childerhouse et al., “Information Flow in Automotive SupplyChains – Present Industrial Practice”, *Industrial Management & DataSystems*, 103(3):137–149, 2003

M.G. Christopher et al. *Supply Chain Vulnerability*, Report for Department of Transport, Local Government and the Regions, Cranfield University, 2002

H. Lee, The triple – a supply chain. *Harvard Business Review*, 102–112, 2004

J. Hallikas, I. Karvonen, U. Pulkkinen, V.M. Virolainen, M. Tuominen, Risk management processes in supplier networks. *International Journal of Production Economics* 90(1):47–58, 2004

Tang CS , Perspectives in supply chain risk management. *Int. J. Production Economics* 103:451–488, 2006

MIX
Papier aus verantwortungsvollen Quellen
Paper from responsible sources
FSC® C105338

If you have any concerns about our products,
you can contact us on
ProductSafety@springernature.com

In case Publisher is established outside the EU,
the EU authorized representative is:
Springer Nature Customer Service Center GmbH
Europaplatz 3, 69115 Heidelberg, Germany

Printed by Libri Plureos GmbH
in Hamburg, Germany